New Perspectives on Chinese Politics and Society

Series Editor

Yang Zhong, Shanghai Jiao Tong University, Shanghai, China

Rapid growth has posed new challenges for sustainable political and economic development in China. This series is dedicated to the study of modern Chinese politics and society, drawing on case studies, field work, surveys, and quantitative analysis. In addition to its empirical focus, this series will endeavour to provide unique perspectives and insights by publishing research from scholars based in China and the region. Forthcoming titles in this series will cover political culture, civil society, political economy and governance.

Yani Wu

# Possibilities and Limitations of Chinese Eco-City Development

palgrave
macmillan

Yani Wu
Zhejiang University and Zhejiang
Urban Governance Studies Centre
(Hangzhou International Urbanology
Research Centre)
Hangzhou, China

ISSN 2634-5404 ISSN 2634-5412 (electronic)
New Perspectives on Chinese Politics and Society
ISBN 978-981-95-1957-6 ISBN 978-981-95-1958-3 (eBook)
https://doi.org/10.1007/978-981-95-1958-3



This Palgrave Macmillan imprint is published by the registered company Springer Nature Singapore Pte Ltd.
The registered company address is: 152 Beach Road, #21-01/04 Gateway East, Singapore 189721, Singapore

# Preface

In an age when rapid urbanization collides with the urgent need for environmental sustainability, eco-city stands as a vision for the ideal human settlement. China, with its development goals to achieve carbon peak by 2030 and carbon neutrality by 2060, has emerged as a global leader in this endeavour, putting significant political will and economic resources into newly built eco-city projects. Yet, amidst the focus on cutting-edge technologies like artificial intelligence and low-carbon technologies, two questions captivated me before writing this book—how can eco-cities not only sustain the planet but also nurture the lives of the people within them? And how does eco-city development not become to be the technological showcase?

This book explores such question, delving into the humanistic dimension of Chinese eco-city planning through the lens of public space design. My fascination with this topic was born from observing urban development for years. Cities, I realize, are more than concrete and steel, they are living ecosystems where people connect and thrive.

This book contends Chinese eco-city planning is not merely a technological or economic pursuit, but a humanistic one that crafts sustainable and liveable habitats. By integrating theories across different disciplines, it proposes considering the interactive relationship between people and environment during the design process can elevate eco-city planning towards a more human-centric path. In a world dominated by digital innovation and environmental metrics, this research could serve as a reminder that the ultimate measure of a city's success lies in the life quality it provides to people.

As you turn these pages, I invite you to explore the humanistic planning concept in the context of disciplinary integration at the theoretical level about how public spaces shape our urban lives, and to consider the potential of humanistic planning to create cities that are not only sustainable but also deeply resonant with the people who call them home at the practical level. I hope this book sparks reflection and dialogue about the future of our cities, where technology and humanity can coexist in harmony.

Hangzhou, China Yani Wu

*Acknowledgements* I would like to express my appreciations to all people who generously give their help and support on this research.

I would give my sincere gratitude to Professor Yang Zhong for giving this publication an opportunity. Thanks to all staffs at the *School of Geography and Planning*, especially Dr. Patricia Aelbrecht, Professor Mara Miele, Professor Oleg Golubchikov and Mrs. Ruth Leo, who were always supportive during my PhD studying in *Cardiff University*.

I sincerely appreciate the support from my parents, Mr. Fengliang Wu and Mrs. Qian Zhang; they warmly and unconditionally supported me in my ten-year studying period in UK. Their forever optimism, gratitude and fortitude towards life have undoubtedly affected my life and made me never think about giving up from beginning to end for urban planning discipline.

Special thanks to my best friends, Dr. Minyao Tang, Miss Yatao Zhao, Miss Yiyan Chen and Mrs. Yujia Wang, they all help me to rebuild the most comfortable state of mind for writing, and to make me rethink the true meaning of faith, hope and love.

Also, I appreciate the help from Mr. Wenjie Chen from *Zhejiang Development and Planning Institute* as well as Professor Jinqun Wu from *Zhejiang University*, their support has made me progress a lot since I came to Hangzhou.

This research is also a gift for my grandmother, who passed away during my writing-up period. Her optimism, perseverance and not giving up on life during the coma in hospital motivated me the most to persist in writing during tough times.

*Competing Interests* The author has no competing interests to declare that are relevant to the content of this manuscript.

# Contents

# About the Author

**Yani Wu** is a postdoctoral researcher at *School of Public Affairs* at *Zhejiang University* and also a postdoctoral researcher at *Zhejiang Urban Governance Studies Centre (Hangzhou International Urbanology Research Centre)*. She obtained the Bachelor's degree in *Urban Studies* and Master's degree in *International Development and Planning* from the *University of Sheffield*, followed by the PhD at *Cardiff University*. Her research interests focus on urban governance, especially Chinese new-type urbanization development, with expertise in analysis of planning policies as well as comparative studies of planning theory and practice. Over the past decade, she interned at *Development Research Centre of the State Council* and *Zhejiang Development and Planning Institute*, participating in planning projects such as *Planning Construction of Jiaxing as a Strategic Hub for Comprehensive Integration with Shanghai* and *Pilot Study on County-Level Urbanization Reform in Zhejiang Province*, which reflects her practical working experience in policy analysis.

# Abbreviations

| | |
|---|---|
| CCICED | China Council for International Cooperation on Environment and Development |
| IISD | International Institute for Sustainable Development |
| KPIs | Key Performance Indicators |
| MEP | Ministry of Environmental Protection |
| MoHURD | Ministry of Housing and Urban-Rural Development |
| NBS | National Bureau of Statistics |
| NDRC | National Development and Reform Commission |
| NGO | Non-Governmental Organization |
| PSPL | Public Space and Public Life |
| RTPI | Royal Town Planning Institute |
| SSTEC | Sino-Singapore Tianjin Eco-City |
| STETC | Singapore-Tianjin Economic and Trade Council |
| UN DESA | United Nations, Department of Economic and Social Affairs |
| WCED | World Commission on Environment and Development |

# List of Figures

# List of Tables

CHAPTER 1

# Eco-City Development: Context, Concepts and Research Design

## 1.1 Introduction

When the extensive economic development model faces profound transformation due to unsustainability, corresponding urbanization model seems to have inevitably undergone a profound transformation, attaching importance to people's living environment and their living quality. Focusing on the current urbanization process, China's long-standing development pattern formed in the traditional industrialization model requires to be reshaped due to the tactics of ecological civilization, so as to promote the economic growth and to realize social high-quality development. Especially in the post-pandemic era, how to achieve green transformation for cities and towns as well as how to realize ecological civilization based on people's ever-growing demand for a better living environment in the true sense have become two major strategic challenges facing Chinese long-term urbanization development. Under social and economic reform background, profound changes have happened to Chinese eco-city development, such as people's social activities, industrial transformations or eco-city's organizational logic.

The long-existing urbanization model in China could be seen as a product of traditional industrial era, while green urbanization is a leap from the traditional state to the new one. This change gives the breeding ground to the rapid development of Chinese eco-cities, or the so-called



Y. Wu, *Possibilities and Limitations of Chinese Eco-City Development*, New Perspectives on Chinese Politics and Society,
https://doi.org/10.1007/978-981-95-1958-3_1

continuous driving force for eco-city's development. It is not easy to judge the feasibility of green urbanization based on whether there are enough successful planning practices of green urbanization under the current development model. However, green development, the foundation of green urbanization, has been regarded as a profound change in the development paradigm; correspondingly, the development mode of eco-city also needs to be changed. This research regards urban environmental problem as an issue of development model, not just an urban problem occurred during the rapid urbanization process. The central point of this research, people-oriented concept, is on initiatives in which Chinese policy-makers contribute to promoting as the primary planning and design goal for urban construction. As the humanistic planning concept has become increasingly mainstreamed for eco-city development, whether and to what extent people-oriented principle has been implemented to Chinese eco-city planning is worthy to be researched.

## 1.2 Research Background: Redefining Chinese Urbanism

### *1.2.1 From Traditional Urbanization to Green Urbanization*

This section aims to focus on the transformative trend of Chinese urbanization from the traditional pattern to the green one, with the purpose of articulating research background to understand how the change of development mode becomes an important reason for the boom of eco-cities.

Urbanization has long been seen as an important driving force for economic development. Generally, city in the modern sense is based upon the industrialization model formed after the industrial revolution. The agglomeration process of population and economic activities in the urban area, known as the urbanization development, has greatly accelerated the industrialization procedure. Human society thus forms the modern social structure according to the industrial civilization as well as the basic urban–rural economic geography characterized by urban-industry and rural-agriculture. The long-existing development model of Chinese urbanization, relating to the economic content carried by specific organization form of city itself, has been largely based upon the logic of traditional industrialization. It cannot be denied that such development model brings progress to mankind, however, it causes a series of unsustainable development issues, especially environmental problems.

In the past few decades, China has undergone a rapid urbanization process as the proportion of its population living in the urban area dramatically increases, making Chinese urban trajectory been characterized by the rapid growth and transformation. In 2019, China's urbanization rate reached 60.6%, and it is estimated appropriately 70% of China's population will live in urban areas. The *National Population Development Plan (2016–2030)* predicts China's total population will be expected to reach 1.45 billion by 2030, and the urbanization rate will be 70% at that time. As is known, among various problems occurring in modern cities, environmental governance has been regarded as the most consensus urban issue to be dealt with. Research conducted by *China Council for International Cooperation on Environment and Development (CCICED)* shows, development model formed during the traditional industrialization era and the urbanization model therefrom, although greatly accumulated industrial wealth, inevitably bring negative impacts to both society and the environment (Liu & Zhang, 2017). At present, how existing cities formed in the era of traditional industrialization become sustainable through green transformation to inject vigour into the economy will become future's development direction of Chinese urbanization.

People naturally treat environmental problems as urban issues rather than development ones, thus, discussion about urbanization referring to the *Global East* mainly emphasizes on how to make existing cities green and sustainable. Traditional urbanization is more like *building parks in existing cities*, while green urbanization seems to be *building cities within natural parks*. As is known, Chinese urbanization experiences several development stages since the reform and opening policies. Key themes have emerged at different timelines during the period of 'hyper-urbanization' (He & Qian, 2017), ranging from the realization of world city version (Wu, 2000) to the boom of projects taking sustainable development as the main planning goal in the mid-2000s onwards (Wu, 2012). From the early 2010s onwards, concepts of low-carbon eco-city have rapidly emerged as hot topics for thinking about the development direction of Chinese future's urbanization.

As discussed above, due to the unprecedented scale of urbanization and corresponding environmental pressures, China has experienced a critical period to tackle ecological issues and has made efforts to meet people's ever-growing needs towards more sustainable and liveable environment. Under such social background, the national government has adopted the humanistic concept, making most of cities being a global trend in which considerations of low-carbon development and people-centred planning principle have been widely taken into urban construction. In this research, eco-city development with the humanistic viewpoint tends to be a more natural and foresighted approach to urbanization planning. The humanistic planning approach in Chinese context pays more attention to solve the contradiction between people and place. In other words, people-oriented planning concept relating to environmental issues might be the traditional concept of *unity of man and nature* (天人合一) in Chinese history.

Regarding Chinese eco-city practices, newly built eco-cities, such as Sino-Singapore Tianjin Eco-City and Caofeidian Eco-City, intend to provide sustainable living environment for people and aim to meet residents' living needs by formulating Key Performance Indicators (KPIs). However, such '*KPI-Driven*' development pattern mainly supported by technological innovations has shown a mode of passive top-down eco-city's planning approach—people's needs are driven by sustainable quantitative indicators formulated by the planning authority, as a result, people's behaviours to a large extent are limited to the fixed indicator index in two ways. On the one hand, while Chinese eco-cities encourage people's behaviours through sustainable quantitative indicators, there still have potential downsides such as individual's use preferences are suppressed in a certain scene—the use of private cars need to be reduced in order to meet the public transport utilization rate, even though residents have their own preferences for choosing cars as their long-distance commute. On the other hand, balancing the need for sustainable practices with residents' choices is a continuous challenge for eco-city's policy-makers, which means urban governance and planning frameworks usually understand these behaviours in the eyes of policy-makers. In addition, eco-city's design stage has obviously provided an opportunity to incorporate urban sustainability goals by assessing corresponding indicators, later stages, such as the implementation and in-use phases, co-determine how effectively and successfully indicator targets are met. As Joss et al. (2012) criticized, 'poor implementation practice due to lack of training or untested technology, and lack of adherence by users may render urban sustainability features partially defunct and lead to underperformance in comparison with the original indicator target' (p. 116).

In assessing contemporary Chinese eco-city development, Caprotti (2014) highlighted the perceived implementation gap between planning targets and actual achievements by using the SSTEC case concerning social sustainability, including public housing, green transport and biodiversity. Also, Jong et al. (2016) illustrated, 'if even SSTEC, the national flagship among the growing number of Chinese eco-city initiatives prompts debates [about implementation evaluation], then questions about implementation success *vs* failure is expected to be centrally important across the spectrum of contemporary Chinese eco-city initiatives' (p. 32). In response, as the *Beautiful China: Eco-City Indicators Guidebook* formulated by *Bluepath City Consulting* (2013) demonstrated, some Chinese commentators assert the adoption of eco-city initiatives as well as related indicator framework is a major step forward, thus, such development represented by the SSTEC project plays significant exemplary role in promoting Chinese sustainable *New-Type Urbanization*.

### *1.2.2 Current Theories and Factors Effecting Green Urbanization*

Since human society transitions from the industrial era to an era characterized by digitalization and sustainability, core determinants of the urbanization model have been experiencing significant transformations, where rapid socio-economic and technological developments are reshaping urban systems. This evolution necessitates a critical examination of the interplay among digital innovation, environmental sustainability and socio-economic dynamics in redefining urban development frame works.

Even though green urbanization or ecological urbanization is a hot topic at the moment in China, in general, the majority of current academic research is still based upon the analytical framework of traditional industrialization pattern, making the understanding of sustainable urbanization being divided into two popular perspectives—the industrialization development procedure and ecological capacity of green urbanization.

On the one hand, understanding of social issues under traditional industrialization thinking mode believes city represents opportunity. 'Population concentration is conducive to economies of scale and technological innovation, so the larger city is, the better it would be' (Zhang, 2021,

p. 5). Unsustainable urban issues, particularly environmental pollution in cities, could be tackled with technological advances. Also, mainstream urbanization scholars, especially economists, hold such viewpoint (Bettencourt, 2013; Fujita and Krugman, 1995 Fuller & Romer, 2014; Glaeser, 2012; Lobo et al., 2020). Lu Ming (2017), the famous Chinese scholar in the field of economic development and urban governance, presented his idea about the size of city—even though urban issues arise due to the large size of cities, resolutions of these problems may also depend on the expansion of cities. However, under traditional industrialization thinking mode, such viewpoint fails to realize cities' unsustainable development issues are mainly caused by the unsustainable development model. On the other hand, emphasizing ecological capacity of green urbanization believes urban development or urban scale should be planned scientifically in accordance with the environmental carrying capacity. Specifically, Song et al. (2019) believes it is unrealistic for any city's development to exceed its environmental carrying capacity.

The traditional thinking about industrialization development is essentially as important as green urbanization since economic development indeed replies on technological breakthroughs in the contemporary society. As a result, understanding of Chinese sustainable urbanization under the above-mentioned two popular aspects may give a new interpretive perspective on eco-city's concept. Specifically, in the context of Chinese eco-city's development, governance mechanism based upon humanistic planning concept is the driving force of development, green and low-carbon technology is the way to achieve development and creating the liveable environment for people is the goal of development.

Zhang (2021) identifies three key factors that determine the urbanization model (See Fig. 1.1), including 'the change in transaction efficiency, the change in the provision of public facilities, and the change in the content of development' (p. 7). Among these factors, development content, including content of production, consumption and trading, has shifted from the industrial wealth characterized by high carbon emission and serious environment damage to high-quality service industries which counts on the economic basis of green urbanization, namely intangible resources such as knowledge, environment and culture.

In recent years, as the whole society has gradually entered to the green and digital development stage, the above-mentioned three key factors are undergoing dramatic transformation, consequently, current urbanization model experiences corresponding changes in China—'when

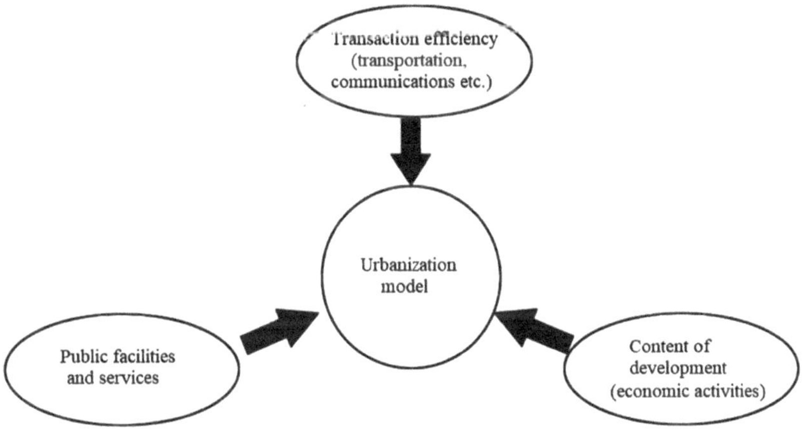


**Fig. 1.1** Three defining factors of urbanization model (*Source* Figure is from Zhang [2021, p. 7])

three defining factors undergo profound changes, requirements for spatial agglomeration of economic development will change, and the content and forms of urbanization will also change accordingly' (Zhang, 2021, pp. 7–8); taking the content and form of social facilities and services as an example. According to the *UN E-Government Survey* 2022, China's e-government key indicators rank top around the world (UN DESA, 2020). In the digital age, new energy and digital reform boomed in the current Chinese society may contribute to making high-quality living achievable at a low cost. Also, the construction of digital platforms contributes to making public services more accessible. The integration of renewable energy systems, such as solar-powered street lighting and electric vehicle charging infrastructure, facilitates green urbanization by mitigating environmental impacts and reducing economic costs. Smart city application like *Shanghai* digital platform provides residents with real-time access to transportation schedules and waste management services.

In the context of ongoing social transformation, the digital technology plays a pivotal role in advancing Chinese eco-city development, with significant implications for enhancing residents' quality of life. On the one hand, the deployment of digital technologies enhances energy efficiency and minimizes energy waste for urban infrastructures. Smart grid systems, for instance, optimize energy distribution, while the integration of renewable energy sources, such as solar and wind power, supports sustainable urban energy frameworks. On the other hand, digital innovations bolster sustainable transportation systems through the adoption of new energy vehicles and intelligent transportation management systems, which collectively reduce carbon emissions and alleviate traffic congestion. These multifaceted applications of digital technology underscore their transformative potential in promoting equitable and resilient urban development within Chinese eco-city development.

## 1.3 Defining the Eco-City Development Under the Global Trend

This section mainly focuses on the evaluation of eco-city's development by tracing the international rise of attention, with the purpose of enabling eco-city concept to be coherently defined as an object of academic study to be explored.

### *1.3.1 Eco-City: Planning Approach to Tackling Environmental Issues*

As a global body of planning practice supported by considerable financial and political resources, the eco-city represents a potent collective pursuit for the sustainable development, that is, the future of human society that urban development should 'not only meet the needs of the present, but also for future's generations to meet their own needs' (WCED, 1987, p. 41). The term *eco-city*, in this research, has been seen as a *symbol* to describe the broader range of contemporary urban-scale initiatives. It aims to contribute to a growing body of critical literatures which plan to identify conceptual and practical issues accompanying the realization of eco-city initiatives.

The emergence of global eco-city concept is an inevitable response to the increasingly industrial-type of urban settlements, where potential environmental issues and social risks may be a threat to living residents, known as the violation of peaceful coexistence between nature and human beings—high concentration of residents and urban infrastructure that distinguish urban settlements leads to pollution and degradation of natural environment. Also, urban residents who are in an artificial environment themselves lose skills to coexist with the natural environment, which has a harmful effect not only on health, but also the psyche of people (Yurzinov, 2014, p. 71). In order to better tackle problems about global environmental crisis on human lives, it might be quite expected a new leading development direction for urban areas. Thus, eco-city emerges, as it is not intended to capture the established set of development goals, rather, it has covered the combination of sustainable concepts and planning practices which attempt to respond a series of contemporary urbanization issues.

The concept of eco-city is widely interpreted in both conceptualization and operationalization—a variety of definitions indeed exist according to the context where it is represented by form of urban initiatives and planning projects. Broadly, an eco-city is 'a human settlement which emphasizes the self-sustaining resilient structure and function of natural environment. It aims to provide [people] the healthy and liveable environment' (Bibri, 2020a, 2020b, p. 7). Roseland (1997) argued there is no single accepted definition for eco-city itself, each one had its own meaning and development characterizes, nevertheless, all reflecting the main planning purpose of eco-city to pursue better living environment for people. Joss (2010) listed the conceptual diversity by demonstrating different global planning projects, and also questioned the usefulness of attempting to capture a narrow form of definition as it is difficult to make an exact definition. However, he defines the eco-city concept based on three analytical categories, namely a substantial scale, occurring across multiple domains, and supported by policy process (Joss, 2011). The conceptualization of the eco-city, in this research, is grounded in Jabareen's (2006) interpretation, which defines it as an overarching metaphor encapsulating diverse urban-ecological strategies aimed at achieving urban sustainability. Such interpretation integrates a broad spectrum of environmental, social and institutional policies designed to manage urban governance effectively (Jabareen, 2006, p. 47), and this research delineates the eco-city as a multi-dimensional construct that aligns urban planning with ecological and social imperatives.

### *1.3.2 Eco-City as an Entity for Human Settlement*

Due to the disagreement over definition of its concept, academic discussion about the eco-city is revolved around wider socio-cultural context related to the form of planning projects and related initiatives, which has been raised issues about 'what an eco-city is, which of the underlying design and strategies contribute more to environmental dimension of sustainability, and the extent to which it considers three dimensions of sustainability' (Bibri, 2020a, 2020b, p. 7). The global ideological trend of eco-city's development, being divided into three main stages shown as follows, intends to argue even though there is no single solution for making living environment sustainable, it is better to rethink the eco-city as an objective that can be achieved in multiple ways to become an ideal model for human settlement.

### 1. **Stage One: Pioneers of the Eco-City**

Joss (2010) summarized a preliminary eco-city phase preceding the *1992 United Nations Earth Summit* in Rio de Janeiro. Rapoport (2010) identified this early period with reference to eco-villages, echoing Barton and Kleiner (2009), who described this as the settlement dating back to the 1970s, and taking the form of small-scale new communities founded on planning principles of self-sufficiency and environmental preservation. In this social context, Register's idea was promoted over the next decade by *Urban Ecology*, an NGO founded in 1975—his idea about the eco-city tended to be much more environmental-focused philosophy, emphasizing mainly on the ecologically healthy city, the biodiversity and the density.

> An eco-city is an ecologically healthy city. No such city exists. There are bits and pieces of the eco-city scattered about in present day cities and through history, but the concept – and hopefully the reality – is just beginning to germinate. (Register, 1987, p. 3).

### 2. **Stage Two: Rio and Beyond (1992–2005)**

Next stage of eco-city development was marked by the mainstreaming of green politics during the 1980s in Northern Europe. Meanwhile, environmental issues mainly concentrated on the international equality, such as the degradation of rural areas (Pepper, 1996) and the growth of squatter settlements (Pugh, 1996) in developing countries. This trend has set the stage for international notion of sustainable development, focusing on discussing around 'three pillars of economy, environment and society' (Kates et al., 2005). Under the trend of sustainable development, the ever-growing appeal of city-level implementation in both North America and Europe was evidenced by the boom of various Asian eco-cities from the beginning of 2000s, such as Songdo International Business District (South Korea), Tajimi (Japan) and Suzhou Industrial Park (China) as well as the announcement of six pilot eco-city schemes by the *Indian Ministry of Environment and Forests* (Joss et al., 2013).

3. **Stage Three: 2005 Onwards**

Amid the globalizing trend, the eco-city concept has gained prominence in shaping governmental urban policies, as evidenced by various international initiatives (Joss, 2011). Notable examples include the UK's four eco-towns (ODPM, 2005), France's nationwide programme ÉcoCités (Joss et al., 2022) proposed in the *National Urban Sustainability Plan* (MEEDDAT, 2008), as well as Japanese government's selection of 13 eco-model cities (JETRO). In addition, the World Bank's *Eco$^2$-Cities* programme offers a broad platform that provides 'practical, scalable, analytical and operational support for cities in developing countries to achieve ecological and economic sustainability' (World Bank, 2010, p. 4). These planning policies have reflected a global commitment to embedding ecological principles within urban governance to address challenges of sustainable development.

It is worth mentioning that given the appropriate funding and great political will, the reification of normative vision of urban futures which focuses on experimental technologies might be relatively easy to achieve in countries with less democratic traditions. Nevertheless, questions still arise about whether these cities will be able to adapt to ongoing technological changes in the future, and to what extent citizen groups that have little say in planning will compliantly adopt these technologies (Caprotti, 2015). At the same time, implementation frictions may be more pronounced in countries with a history of public participation in planning processes. For instance, eco-city initiatives in UK have met with widespread local opposition. In the proposed *Whitehill-Boden* development, campaigners complained that public input was ignored by the political and business actors involved (Boden Area Action Group, Undated), describing the scheme as 'a politically driven process, and local community positively rejects because it threatens their values and their way of life' (Bordon Area Action Group, 2009). Eco-city initiatives are vulnerable to changes in government—after 2010 national elections, the policy framework was abandoned and local projects made little progress (Tomozeiu & Joss, 2014).

The proliferation of sustainability and ecological modernization discourses has shaped the historical evolution of eco-city development, revealing a series of transformative shifts that situate the eco-city within broader social and political context, and such transformation is intricately linked to the progression of environmental thought since the 1960s. However, the absence of a universally accepted definition complicates eco-city's conceptualization, particularly as it intersects with related paradigms of sustainable urban development, such as smart city, healthy city or liveable city. This convergence prompts critical inquiries into the distinct identity and theoretical boundaries of the eco-city as a subject of study. Thus, to avoid overlapping other vision of human settlements by a wide variety of eye-catching labels, it is vital for the eco-city to construct its own field of enquiry. In recent years, a new wave of mainstreamed eco-city has been paid much attention on the form of technological showcase—primarily serving commercial, political and institutional goals, but whose social dimensions remain ignored. It may have negative effects on its ability about implementation in human settlement (Carvalho, 2014; Shwayri, 2013; Yigitcanlar & Lee, 2014). Such emerging trend to some extent makes this research being meaningful as it focuses on the ignored social sustainability aspect of eco-city's development, that is, the perspective of people-oriented eco-city planning.

## 1.4 Research Aim, Research Questions and Research Contributions

### *1.4.1 Research Aim*

This research aims to address the *global conceptual-Chinese practical* gap in knowledge, by critically identifying whether there are any evidences to show the people-oriented planning concept rooted in the universal theoretical perspective has happened in Chinese eco-city development, if yes, in what ways and to what extent. As is known, in the global context, the traditional humanistic concept relating to environmental issues within urban planning discipline mainly focuses on the participatory planning or collaborative planning. While in Chinese planning practices, the appropriate theory of people-oriented concept relating to environmental issues might be the time-honoured concept of *unity of man and nature* (天人合

—), which mainly drives Chinese eco-city development. Further analysis about such differences will be discussed in the next chapter.

This research plans to adopt the interdisciplinary approach to develop a multi-dimensional conceptual framework about *people–place interactive relationship*, which is seen as the manifestation of *humanistic planning concept under the idea of ecology* in this research. It uses the typical qualitative methodological approach under the framework of three-layer research design with a single case study, drawing on literatures about three different disciplines (urban planning, urban design and applied psychology) to conceptualize the syncretic match among people, place and ecology, as well as dynamics of responses to the humanistic change. And the main aim of this research is shown as follows.

> To investigate the concrete meaning and feasibility of the global mainstream people-oriented concept (theory) into Chinese eco-city planning development (practice), including the core connotation and implications in a broader sense of humanistic planning under the context of ecology. To realize this conceptual exploration, it mainly focuses on the newly built eco-city through the prism of residents' public lives (esp. their daily outdoor activities) in public spaces.

### 1.4.2 Research Questions Related to Each Research Objective

To achieve the main research aim, four objectives have been set out.

**Objective One**: to review theories relating to the *humanistic planning under the idea of ecology* from three different disciplines and to explore the interaction among these discourses, for building an analytical framework responding to the research gap, while also offering the theoretical support for further research analysis. Also, to review the new-type conceptual model related to human-centred design for corroborating the concrete meaning of people-oriented concept, namely considering people's needs and behaviours.

To meet the first objective, it requires a broad exploring about literatures concentrating on both humanistic and ecological aspects—*Garden City*, as a way to provide the classic theoretical foundation for people–place relationship that supports eco-city's planning and design towards the humanistic direction (Dobbins, 2009; Parham, 2013). *Ecological Urbanism*, as the emerging humanistic design concept since the turn of twenty-first century, aims to seek for an approach which can have the capacity to ease inherent conflictual conditions between ecology and urbanism, and helps to set out guidelines to identify human-centred design elements (Hodson & Marvin, 2010; Spirn, 2000). *Environmental Psychology*, one of the humanistic concepts outside urban planning and design disciplines, refers to a 'pre-paradigmatic stage' (Lang et al., 1974, p. 11) that focuses on the transaction between people and the physical environment, by exploring people's psychological feelings, seen as people's 'invisible needs' in this research. This objective is achieved by a critical review on the related literature in Chapter 2, and also a review on the new-type conceptual model to demonstrate why people's needs and behaviours matter in Chapter 3.

**Objective Two**: to review core influencing factors for supporting Chinese eco-city development, and to review how Chinese urban development moves to the new development stage by taking the humanistic planning concept into consideration.

**Research Question One: Why people-oriented eco-city development pattern should be regarded as the mainstream planning approach to current Chinese urbanization development path?**

To meet the second objective, Chapter 4 firstly delineates some general information about the SSTEC project, including its geographical and political conditions, its bilateral supervisory management framework, which are basic but unique characteristics that can affect further data analysis of this research by demonstrating its project positioning. Then, in this chapter, it focuses on how people-oriented concept reflects in Chinese eco-city project. Specially, from the perspective of urban planning, indicator system of the SSTEC has shown the ongoing trend towards people-oriented development pattern since its initial planning and later construction processes. From the perspective of urban design, the masterplan has been formulated to provide a clear blueprint to guide the SSTEC's overall development to achieve its KPIs, and also to offer people a life-convenient living settlement under the general people-oriented design concept. In the scope of second half of this chapter, it plans to introduce elements that could constitute responsive public spaces of the SSTEC, including permeability and legibility as well as variety, to argue people-oriented concept has been rooted in the level of concrete urban design approach.

**Objective Three**: to explore the public life in public spaces from the perspective of people's experience, and to identify key findings by exploring existing obstacles that may limit Chinese eco-city project as well as existing planning initiatives to meet the goal about considering people's needs during the actual planning and design processes. It also aims to explore potential influences about how people–place interactive relationship is to be applied into eco-city's place-making process.

**Research Question Two: How Chinese eco-city planning regulations and proposed projects consider people-oriented concept for both current and future eco-city development?**

- With the advancement of Chinese eco-city planning practices, in what aspects has the people-oriented concept been improved?
- How can insights and knowledge from current research, and also from the analysis of the empirical data for this research be used to provide potential innovative approaches to Chinese eco-city development?

The first half of Chapter 5 discusses reasons for choosing four research sites within the SSTEC to narrow down the analytical scope of people–place relationship. Then, it introduces the evaluation about people's activities to lay the groundwork for further analysis about how people measure good and poor spatial environment by their behaviours. In the scope of second half of this chapter, as Chapter 3 discusses the importance of people's needs by using a new-type conceptual model to understand the human-centred design in a historical context, Chapter 5 has identified what kinds of people's needs should eco-city meet, both physically and psychologically. Based upon research findings, sense of safety (psychological aspect) and social infrastructure (physical aspect) have been discussed as two forms of the behavioural connection between people and place, which has been deemed as two important leverage points of eco-city's place-making process in the next chapter. All research findings are used as the conclusive basis for the discussion in Chapter 7.

Chapter 6 aims to respond to the original intention of doing this research—identifying limitations and possibilities of current Chinese eco-city development in order to make its planning and design is more humanistic. Thus, in this chapter, it critically explores potential constraints on promoting the behavioural connection between people and place, to respond to reasons for the existence of conceptual-practical gap identified in the previous chapters. In addition, by discussing the general information about existing traditional eco-city's design process, possible improved design process in terms of recommendation for practice has been put forward to demonstrate the implication of behavioural connection between people and place in the eco-city planning development.

### *1.4.3 Research Contributions*

This study investigates the humanistic planning through an empirically grounded case study, contributing to the discourse on people-oriented planning concept within Chinese eco-city development. It offers novel theoretical insights, complemented by empirical findings, to advance the understanding of urban sustainability framework. Empirically, this research offers an original analysis of the conceptualization of people–place relationship. Also, it provides an analysis of the humanistic planning concept, emphasizing its conceptualization of the people–place relationship.

Drawing on primary findings from the Sino-Singapore Tianjin Eco-City case study, this research addresses a critical gap in literatures concerning social sustainability in the eco-city development framework, an area that has received limited scholarly attention. And to fill such research gap, this research posits eco-city development as the continuous driving force for Chinese new-type urbanization, aligning with current development goals, such as carbon peaking and carbon neutrality as well as digital reform. It considers the fundamental development principle of Chinese eco-city that lies in the humanistic concept, with low-carbon and digital technologies serving as means for sustainable development. Ultimately, providing the liveable environment for residents is the paramount development objective. This nuanced humanistic exploration contributes to academic debates and also holds implications for policy and practice, particularly in the dynamic of contemporary urban development. Detailed explanation is as follows.

This research establishes connections between theory of applied psychology and theories of urban planning and design, building on discussions about the humanistic planning concept by paying attention to people's psychological needs. Positioned as an interdisciplinary study, it enriches eco-city literatures and humanistic planning practices. Moreover, this research provides Chinese practical response to the global concept of eco-city development, using the SSTEC case study as a focal point. It emphasizes a trinity of economic-environmental-humanistic urban governance framework, distinguishing Chinese eco-city's development pattern from other countries' models. In other words, during the current development era of green urbanization, this research believes Chinese eco-city development exemplified by the SSTEC aligns with current national planning strategic about green low-carbon development, transitioning from the traditional urbanization which focuses on economic and environmental aspects to the economic-environmental-humanistic governance approach.

## 1.5 Research Structure

This chapter redefines Chinese urbanism from the traditional development pattern to green urbanization to understand urban China. It gives the general overview about the global trend of eco-city development to rethink the eco-city as an objective that can be achieved in multiple ways to become an ideal model for human settlement, and it also explicates research aim, research questions and some research contributions.

Chapter 2 draws on a different set of literatures (related to people–place relationship, people-oriented concept and ecology) to argue that eco-city planning ought to be seen as the humanistic planning. Specially, it illustrates three different theories related to humanistic planning under the context of ecology, and then establishes the analytical framework for this research.

Chapter 3 introduces the new-type conceptual model of human-centred design to argue why people's needs and behaviours matter.

Chapters 4 and 5 aim to analyse the empirical research data about people–place relationship to measure the conceptual-practical gap, from the perspective of people's experience as well as from the perspective of planning and design process.

Chapter 6 aims to explore key findings for identifying some obstacles that limit Chinese eco-city project to meet the goal about considering people's needs during the actual planning and design process, and also to explore potential influences of people–place interactive relationship to be applied into eco-city's place-making process.

Chapter 7 summarizes both findings and arguments. It also concludes research limitations and research contributions.

## 1.6 Overview of Methodological Choice

This research situates in *the humanistic planning under the context of ecology*, an applied field that requires insights from multiple disciplines—urban planning, urban design and psychology respectively. In a narrow sense, it intends to explore people's behaviours in public spaces within the eco-city, in order to search for the interaction between people and place. From the perspective of research aim listed in this chapter, it requires qualitative data and suggests the necessity of researcher's involvement as well as the (participatory) observation. In a broad sense, this research is expected to discuss some specific challenges, issues or problems of humanistic planning and to deepen (and also broaden to a small extent) the implication of people–place interactive relationship in Chinese eco-city development. As qualitative method focuses on the in-depth validity over generalizability, data collection and analysis used by qualitative techniques are drawn from particular context in this research.

This research has adopted a holistic-view research approach, that is, both inductive (data-driven) and deductive (theory-driven) analysis (Braun et al., 2019, p. 297) are combined to interpret qualitative data collected from the case study. The reason for this choice is to maximally find out the match (applicability) between global universal theories and eastern practices for ensuring that the result of this study is trustworthy and meaningful, as there is a certain degree of difference between global universal theories in guiding eastern practices. Specifically, *the humanistic planning concept under the ecology context* has been seen as the focal point of this research. In Chinese context, the appropriate theory might be the traditional concept of *the unity of man and nature* (天人合一), this research aims to learn from this interactive relationship, to see whether there are any evidences to show the use of global universal humanistic planning and design concept in Chinese eco-city development, if yes, in what ways and to what extent. One the one hand, from the perspective of inductive analysis, such analytical method can capture both manifest and latent meanings (Braun et al., 2019, p. 851), as it not only summarizes and reorganizes observational diaries, questionnaires and interview transcripts, more importantly, it indeed contributes to connecting research objectives with research questions to identify and to interpret key features of research data. On the other hand, from the perspective of deductive analysis, this research innovatively combines both humanistic and ecological elements from different disciplines, in order to build the fundamental conceptual and analytical framework for guiding the whole research process. Focusing on ***the humanistic planning under the context of ecology, the people–place interactive relationship***, derived from three theories in different disciplines, has served as the core connotation of people-oriented concept, which is used to guide data collection during the fieldwork and data analysis after the fieldwork. Specifically, the *Public Space and Public Life (PSPL) Survey* is used to collect data about residents' outdoor activities (people's behaviours) in public spaces within the eco-city (place) and their attitudes or opinions (people's cognition) on the spatial quality of public spaces within the eco-city (place). Moreover, data discussion has always revolved around the people–place relationship derived from the combined theoretical framework discussed in Chapter 2, to explore both forms and implications of behavioural connection between people and place in the eco-city development, and also to response to *Objective Three*.

## 1.7 The Three-Layer Research Design with a Single Case Study

For a more logical and systematic data collection and analysis, *the three-layer methodological framework with a single case study* has been designed as the research strategy for this research, for exploring research data about the people–place interactive relationship in Chinese eco-city project, and for examining people's needs and behaviours when they are staying in public spaces. A set of qualitative research methods are properly chosen and are used in each data collection procedure. Three influencing factors have been considered for selection of research strategy -the type of research questions, the degree of control over actual behavioural events faced by the researcher and the focus of attention about contemporary phenomena (Yin, 2014).

In this research, the typical qualitative methods are participant observation, semi-structured interviews, qualitative self-completion questionnaire to guarantee 'complete answers to research questions and balanced results' (Johnson et al., 2007, p. 125). It is worth mentioning that data collection under such strategy has been designed as the so-called *observation sandwich* to ensure the reliability of data, and to increase credits of the internal validity. That is, verification becomes the theme of this research strategy—participant observation used during the pilot study is followed by the street-intercept survey, and then they are followed by in-depth interviews, with the single case study design acting as the toothpick of *observation sandwich*.

### 1.7.1 Single Case Study Strategy

Case study refers to a programme or an activity framed in the specific time and place. According to Bhattacherjee (2012), the case study is a suitable choice for research, where the phenomenon during a period of time in one or several sites can be intensively researched. It is 'defined by interest in an individual case, not by methods of inquiry used, and the object of study is a specific, unique and bounded system' (Stake, 2008, p. 123). Case study helps the researcher to understand 'complex, temporal processes rather than factors or causes' (Bhattacherjee, 2012, p. 94), and contributes to providing more contextualized and more authentic interpretation of the phenomenon. Yin (2014) demonstrated both importance and applicability of case study design in the explanatory research of social science. It is functioned as an empirical inquiry that can research contemporary social phenomenon in depth within the real-world context, especially under the situation of unclear boundaries between social phenomenon and context. 'The suitability of case study seen as the subject of the research is a contemporary event, and the researcher has little control over relevant behaviours involved in the event' (pp. 15–16).

The overall aim of this research focuses on the understanding about humanistic issues in Chinese eco-city development, instead of the comparative or general perception about eco-city planning. Specially, this research strategy is designed based upon research questions to show how research data has been collected and analysed from two aspects—on the one hand, how to define the people–place interactive relationship (how do people measure the spatial environment by their behaviours and how does the spatial environment impact people's behaviours), and how this interactive relationship influences eco-city development (whether the design pattern considering people–place relationship can be regarded as the possible improved approach to Chinese eco-city development). On the other hand, what are constraints and initiatives faced by Chinese eco-city development when implementing the humanistic planning concept. Based on the above two points, there have been more than one variable of interest (e.g. people, place and constraints) and multiple sources of needed evidence in this research; the researcher has less control.

Such designed case study strategy makes a contribution to this research, and also to existing literatures, especially due to the lack of discussion about people-oriented planning concept under the context of ecological development in academia. Thus, more exploration and expanding on recent eco-city's academic research need to be explicitly made by a broader humanistic perspective and by the interdisciplinary way of thinking. That is, this research design aims to understand topics and issues of humanistic planning concept from multiple theories in literatures and to refine the people–place relationship from the chosen case study. In addition, referring to the category of case study, Yin (2014) classified three types—exploratory, explanatory and descriptive ones, which are not seen as being separate or being hierarchy. This research is designed as the *exploratory-explanatory* case study to make research findings be more convinced.

As Flyvbjerg (2006) illustrates, the successful choice of paradigmatic case cannot be guaranteed in advance of the study itself. 'There was no assumption, therefore, that the case would, on closer examination, reveal itself to be truly paradigmatic, rather than force a categorisation on the case through which to interpret the findings, a more grounded approach was adopted, whereby the case was expected to be uniquely shaped by its particular context' (Robert, 2016, p. 131). *The Sino-Singapore Tianjin Eco-City (SSTEC)* has been chosen as a case to examine and to refine the analytical framework by exploring some typical issues and concerns about the humanistic planning and design concept (people–place relationship), which does not exist in literatures. Further crucial reason for selecting the SSTEC as the single case study to research because it appears to represent what Flyvbjerg (2006) believes a 'critical case'. As mentioned above, a case study design has been adopted to answer research questions and to collect data for exploring real-world situations about the interactive relationship pattern between people and place within Chinese eco-city. Process of choosing a case study sample is selective with a purpose. Accordingly, a research sample should be selected to investigate the people-oriented planning concept (public life in public spaces) in full details. Thus, choice of a case study sample tends to be the most influential and representative Chinese eco-city project to the application of humanistic planning and design concept. That is, the chosen sample ought to obviously represent the prevailing planning practices and also it may potentially influence the future development of Chinese eco-cities.

Caprotti and Gong (2017) made the case for 'humanizing' newly built eco-cities by emphasizing on urban social sustainability through the lens of experiences of new residents. They argued 'there is a need to focus on the ways in which socially sustainable urban environments are constructed, in new urban spaces, through relational networks comprised by interactions between residents, buildings, facilities and specific (e.g. domestic) spaces' (p. 45). However, as Joss and Molella (2013) revealed in their research, 'Chinese urban experiment as a *techno-city*, putting much emphasis on how the eco-city development is carried out in conjunction with the development of a broader technological project' (p. 121). Moreover, current academic studies on the SSTEC, the first eco-city that developed collaboratively between countries and the experimental and innovative pilot project in China, mainly focus on the descriptive or technological analysis about planning documents and second-hand planning policies. They examine 'whether eco-cities exemplify urban sustainability ideals, discuss challenges of implementing eco-cities, or question how eco-city's experience can be used to develop indices and assessment measure for future projects' (Chang, 2018, p. 62); lacking research focuses on the humanistic aspect and social sustainability. Such research gap in academic literatures has determined the reason for case study selection—combining the humanistic planning and design concept with the SSTEC practice.

### *1.7.2 The First Layer: An Exploratory Pilot Study*

In order to investigate whether crucial components of this research are feasible and can predict an appropriate sample size for the full-scale research, small-scale pilot study with an exploratory case study has been adopted as the first layer of the research design. The most important reason for conducting the pilot study is because it is crucial for the researcher to avoid wasting time and resources that can collect enough data that responds to answering research questions.

The first layer of 'sandwich' is made of desk research and preliminary participant observation by the researcher known as the pilot study. The aim of this stage is to broaden more knowledge on research topics and to explore the initial information about the SSTEC project, checking the sampling and evaluating the research design. During this stage, it aims to narrow down research questions from the analytical framework and to identify how research questions are properly formulated for the application to research methods. Document analysis and preliminary participant observation at this stage respond to *Objective Two* and *Research Question One*, by reviewing core influencing factors for supporting Chinese eco-city development as well as how Chinese urban development move to the new development stage of people-oriented eco-city.

### *1.7.3 The Second Layer: The Multiple-Method Qualitative Data Collection Approach*

It is the most important stage of this research, via the multiple-method data collection to explore the people–place relationship in the real-world eco-city project. It is worth mentioning that typical qualitative research methods including participant observation and questionnaires at this stage respond to *Objective Three* and *Research Question Two*, by exploring the public life in public spaces from the perspective of people's experience to provide the foundation of preconceptions for further data collection.

The second-layer stage, using the *Public Space and Public Life (PSPL) Survey* proposed by Jan Gehl, is the approach utilizing empirical analysis to understand how the built environment can promote well-being and effect social interactions between people. Specially, the *PSPL Survey* adopts further on-site observation and qualitative questionnaires, attempting to verify data collected from the pilot study, as research data of the first layer tends to be occupied by the subjective perception of researcher's feelings and judgement, such as the evaluation about spatial quality of several public spaces within the SSTEC.

On the one hand, this stage of data collection focuses on the public space analysis by means of mapping and on-site counting, aiming to explore types of people's outdoor activity and users of public spaces (e.g. different ages and occupations). In addition, it emphasizes on the public life survey by means of on-site observation, aiming to explore a variety of spatial elements that can potentially stimulate or prevent people to spend more time in public spaces. On the other hand, the street-intercept survey by the qualitative questionnaire, being functioned as the supplement to observation at data collection stage, aims to explore people's more detailed perspectives and objective descriptions about their perception and attitudes (defined as 'invisible' needs) towards public places, as well as their comments on social infrastructures within the eco-city; that is, to investigate and to identify criteria and causes of good or bad spatial qualities in their eyes.

### 1.7.4 *The Third Layer: Data Analysis and Further Data Collection*

The last research stage, a series of in-depth semi-structured interviews with planning practitioners and management operators of the SSTEC project, is based upon the analysis of all research data from the former two layers. It is concerned about answering the last main research question—understanding whether and how does the case study project consider impacts or implications on the interactive pattern between people and place. At the end of this stage, by combining all findings, the overall understanding of humanistic planning under the context of ecological development aims to be achieved, and potential recommendations on people-oriented planning principle and procedure is devised. An explanatory case study, being used to 'do causal investigation and to understand the complex social phenomena' (Yin, 2014, p. 6), is designed to apply in the second-layer research strategy. It is worth mentioning that typical qualitative research methods including participant observation and in-depth interviews at this stage respond to *Objective Three* and *Research Question Two*, by synchronizing key findings to explore the existing obstacles that may limit Chinese eco-city project as well as the existing planning initiatives to meet the development goal about considering people's needs. The process of data collection and analysis is shown in the following Figs. 1.2 and 1.3.

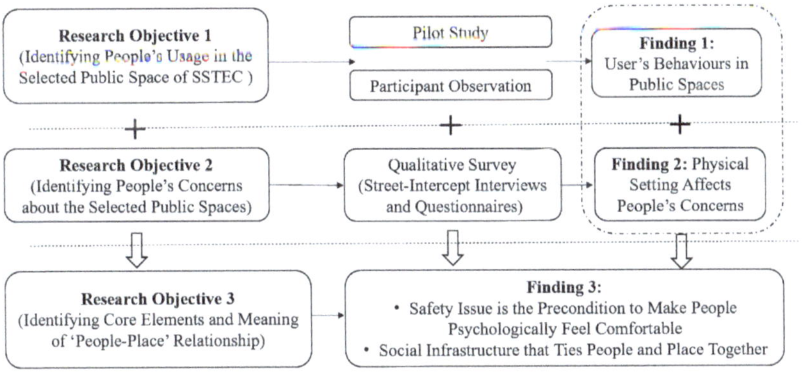


**Fig. 1.2** Research objectives, methods and findings (*Source* Figure is drawn by the researcher)

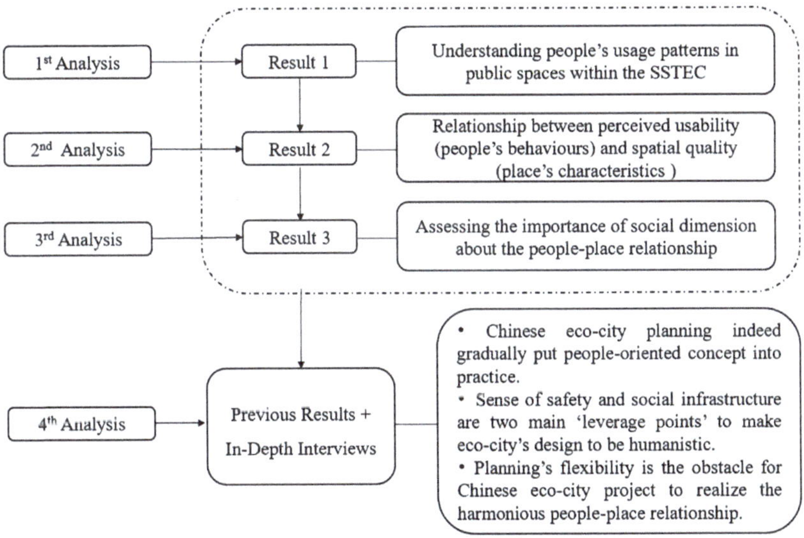


**Fig. 1.3** Overall data analysis (*Source* Figure is drawn by the researcher)

## References


Barton, S., & Pineo, R. (2009). *Human Benefits of Green Spaces.* Cited in: Barton, S., & Pineo, R. (2009). *Sustainable Landscapes Series* (1st ed.). University of Delaware.

Bettencourt, L. M. (2013). The Origins of Scaling in Cities. *Science, 340*(6139), 1438–1441.

Bhattacherjee, A. (2012). *Social Science Research: Principles, Methods, and Practices* (2nd ed.). Global Text Project [Viewed September 22, 2018]. Available at: http://scholarcommons.usf.edu/oa_textbooks/3

Bibri, S. E. (2020a). Smart Eco-City Strategies and Solutions for Sustainability: The Cases of Royal Seaport, Stockholm and Western Harbour, Malmö, Sweden. *Urban Science, 11*(6), 1–42.

Bibri, S. E. (2020b). The Eco-City and Its Core Environmental Dimension of Sustainability: Green Energy Technologies and Their Integration with Data-Drive Smart Solutions. *Energy Informatics, 2020*(3), 1–26.

Bluepath City Consulting. (2013). Beautiful China: Eco-City Indicators Guidebook. Tongxin Press: Beijing. Bordon Area Action Group. (2009). *Government-Council Stitch-Up for Bordon.* (Online). Available at: http://www.baaga.co.uk.. Accessed on April 19, 2023.

Bordon Area Action Group. (Undated). *Welcome to BAAG* (Online). Available at: http://www.baaga.co.uk. Accessed on April 19, 2023.

Braun, V., Clarke, V., Hayfield, N., & Terry, G. (2019). *Thematic Analysis.* Cited in: Braun, V., et al. (2019). *Handbook of Research Methods in Health Social Science* (pp. 843–860).

Caprotti, F. (2014). Eco-Urbanism and the Eco-City or Denying the Right to the City? *Antipode, 46*(5), 1285–1303.

Caprotti, F. (2015). *Eco-Cities and the Transition to Low Carbon Economies* (1st ed.). Palgrave Macmillan.

Caprotti, F., & Gong, Z. Y. (2017). Social Sustainability and Residents' Experiences in a New Chinese Eco-City. *Habitat International, 61*(2017), 45–54.

Carvalho, L. (2014). Smart Cities from Scratch? A Socio-Technical Perspective. *Cambridge Journal of Regions, Economy and Society, 8*(1), 43–60.

Chang, I.-C. C. (2018). *Let's Focus on Connections: Reading Chinese Eco-Cities from a Relational, Multi-Scalar Perspective.* Cited in: Yu, L., & Caprotti, F. (2018). *Sustainable Cities in Asia* (1st ed., pp. 60–74). Routledge.

Dobbins, M. (2009). *Urban Design and People* (1st ed.). John Wiley & Sons Inc.

Flyvbjerg, B. (2006). Five Misunderstandings About Case-Study Research. *Qualitative Inquiry, 12*(2), 219–245.

Fujita, M., & Krugman, P. (1995). When is the Economy Monocentric? *Regional Science and Urban Economics, 25*(4), 505–528.

Fuller, B., & Romer, P. (2014). Urbanization as Opportunity. *World Bank Policy Research Working Paper No.6874* (Online). Available at: https://papers.ssrn.com/sol3/papers.cfm?abstract_id=2439696. Assessed on May 20, 2019.

Glaeser, E. L. (2012). *Triumph of the City: How Our Greatest Invention Makes Us Richer, Smarter, Greener, Healthier and Happier* (1st ed.). Penguin Press.

He, S. J., & Qian, J. X. (2017). From an Emerging Market to a Multifaceted Urban Society: Urban China Studies. *Urban Studies, 54*(4), 827–846.

Hodson, M., & Marvin, S. (2010). Urbanism in the Anthropocene: Ecological Urbanism or Premium Ecological Enclaves? *Cities, 14*(3), 298–313.

Jabareen, Y. R. (2006). Sustainable Urban Forms: Their Typologies, Models and Concepts. *Journal of Planning Education and Research, 26*(1), 38–52.

Johnson, R., et al. (2007). Toward a Definition of Mixed Methods Research. *Journal of Mixed Methods Research, 1*(2), 112–133.

Jong, M. D., et al. (2016). Eco-City Development in China: Addressing the Policy Implementation Challenge. *Journal of Cleaner Production, 134*, 31–41.

Joss, S. (2010). Eco-Cities—A Global Survey 2009. *WIT Transactions on Ecology and the Environment.*

Joss, S. (2011). Eco-Cities: The Mainstreaming of Urban Sustainability: Key Characteristics and Driving Factors. *International Journal of Sustainable Development Planning, 6*(2), 1–18.

Joss, S., & Molella, A. P. (2013). The Eco-City as Urban Technology: Perspective on Caofeidian International Eco-City China. *Journal of Urban Technology, 20*(1), 115–137.

Joss, S., Cowley, R., & Tomozeiu, D. (2013). Towards the 'Ubiquitous Eco-City': An Analysis of the Internationalisation of Eco-City Policy and Practice. *Urban Research and Practice, 6*(1), 54–74.

Joss, S., Tomozeiu, D., & Cowley, R. (2012). Eco-City Indictors: Governance Challenges. *WIT Transactions on Ecology and the Environment, 155*(12), 109–120.

Joss, S., David, H. and Serra, L. (2022). Eco-Neighbourhood and the Question of Locational Advantage: A Socio Spatial Analysis of French 'ÉcoQuartiers'. *Cities, 126*, 103643.

Kates, R., Parris, T., & Leiserowitz, A. (2005). What is Sustainable Development? Goals, Indicators, Values, and Practice? *Environment, 47*(3), 8–21.

Lang, J., Burnette, C., Moleski, W., & Vachon, D. (1974). *Designing for Human Behaviour: Architecture and the Behavioural Sciences* (1st ed.). John Wiley & Sons Inc.

Liu, S. J., & Zhang, Y. S. (2017). *A New Era of Green Development: China's Green Transformation 2050.* Report from China Council for International Cooperation on Environment and Development.

Lobo, J. M., et al. (2020). *Urban Science: Integrated Theory from the First Cities to Sustainable Metropolises* (Online). Available at: https://papers.ssrn.com/sol3/papers.cfm?abstract_id=3526940. Accessed on January 11, 2023.

Lu, M. (2017). *The Power of Space: Geography, Politics and Urban Development (空间的力量:地理、政治与城市发展)* (2nd ed.). Shanghai People's Publishing House.

MEEDDAT. (2008). *ÉcoCités: Cities of Sustainable Development: 15 Major Architectural, Social and Energy Innovation Projects at the City Level.* Ministry of Ecology, Energy, Sustainable Development and Regional Planning (Online). Available at: https://www.ecologie.gouv.fr/. Accessed on March 12, 2019.

ODPM. (2005). Planning Policy Statement 1: Delivering Sustainable Development. London: HMSO.

Parham, S. (2013). *Shaping Sustainable Urbanism: Are Garden Cities Answer?* (Online). Available at: http://uh-sustainable.co.uk/docs/Shaping_sustainable_urbanism.pdf. Accessed on January 20, 2018.

Pepper, D. (1996). *Modern Environmentalism: An Introduction* (1st ed.). Routledge.

Pugh, C. (1996). *Sustainability and Sustainable Cities.* Cited in: Pugh, C. *Sustainability, Environment and Urbanization* (1st ed., pp. 135–177). Earthscan.

Rapoport, E. (2010). *Defining the Sustainable City: The Discourses and Convictions Behind the Eco-City.* Conference Paper Presented at the Summer Symposium on Sustainable Systems, Finland (Online). http://engd-usar.cege.ucl.ac.uk/project/view/idprojects/9#outputs. Accessed on February 15, 2019.

Register, R. (1987). *Eco-City Berkeley: Building Cities for a Healthy Future* (1st ed.). North Atlantic Books.

Robert, C. (2016). *Eco-Cities: Technological Showcases or Public Spaces?* University of Westminster.

Roseland, M. (1997). Dimensions of the Eco-City. *Cities, 14*(4), 197–202.

Shwayri, S. (2013). A Model Korean Ubiquitous Eco-City? The Politics of Making Songdo. *Journal of Urban Technology, 20*(1), 39–55.

Song, F., et al. (2019). Research Progress of Urban Ecological Carrying Capacity. *Environmental Ecology, 2019*(1), 53–61.

Spirn, A. W. (2000). *Ian McHarg, Landscape Architecture and Environmentalism: Ideas and Methods in Context.* Cited in: Conan, M. (Ed.). (2000). *Environmentalism and Landscape Architecture* (pp. 97–114). Dumbarton Oaks.

Stake, R. E. (2008). *Qualitative Case Studies.* Cited in: Denzin, N. K., & Lincoln, Y. S. (2008). *Strategies of Qualitative Inquiry* (3rd ed., pp. 119–149). Sage Publications, Inc.

Tomozeiu, D., & Joss, S. (2014). Adapting Adaptation: The English Eco-Town Initiative as Governance Process. *Ecology and Society, 19*(2), 20.

United Nations, Department of Economic and Social Affairs, Population Division. (2022). *United Nations E-Government Survey* (Online). Available at: https://publicadministration.un.org/egovkb/en-us/Reports/UN-E-Government-Survey-2022. Accessed on January 11, 2023.

WCED. (1987). *Report of the World Commission on Environment and Development: Our Common Future* (Online). Available at: https://sustainabledevelopment.un.org/documents/our-common-future.pdf. Accessed on June 30, 2022.

World Bank. (2010). *Eco2 Cities Brochure* (Online). http://siteresources.worldbank.org/INTEASTASIAPACIFIC/Resources/Eco2_Cities_Brochure. Accessed on February 15, 2019.

Wu, F. L. (2000). The Global and Local Dimensions of Place-Making: Remaking Shanghai as A World City. *Urban Studies, 37*(8), 1359–1377.

Wu, F. L. (2012). China's Eco-Cities. *Geoforum, 43*(2), 169–171.

Yigitcanlar, T., & Lee, S.-H. (2014). Korean Ubiquitous-Eco-City: A Smart-Sustainable Urban Form or a Branding Hoax? *Technological Forecasting & Social Change, 89*, 100–114.

Yin, R. K. (2014). *Case Study Research: Design and Methods* (5th ed.). Safe Publications.

Yurzinov, I. L. (2014). Eco-Town: Current Status and Prospects. *Economy Taxes Right, 6*, 71–73.

Zhang, Y. S. (2021). China's Green Urbanization in the Perspective of Ecological Civilization. *Chinese Journal of Urban and Environmental Studies, 9*(1), 1–17.

CHAPTER 2

# Bridging Knowledge and Action: The Humanistic Planning Foundation for Chinese eco-City Development

## 2.1 Introduction

'Planning is located precisely at the interface between knowledge and action' (Friedmann & Hudson, 1974, p.2). Being a practical discipline with strong applicability, urban planning is a subject that seeks to improve the well-being and people's quality of life, by recognizing that cities are not just physical spaces but also social and cultural environment that shape people's lives. The humanistic planning concept aims to create cities that are more sustainable and humane by recognizing cities are not just physical structures, but also dynamic and complex social systems that indeed require careful attention to people and social dimensions of urban life. At the practical level, the humanistic planning mainly involves working closely with communities to understand people's needs, developing plans and policies that prioritize individuals' interests.

While there have been a variety of ideas presented in the name of humanism, a unifying theme across the variations of humanistic concepts within urban planning discipline is the natural pivot to action in the present world to improve conditions for future generations. Humanistic planning, the concept advocated by the world-famous planner Jane Jacobs, has been the noble cause that she pledged her life to. The premise of humanistic planning is simple, building cities for people, not cars, nor buildings. As Jacobs (1985) stated, 'there is no logic that can be



Y. Wu, *Possibilities and Limitations of Chinese Eco-City Development*, New Perspectives on Chinese Politics and Society,
https://doi.org/10.1007/978-981-95-1958-3_2

superimposed on the city. People make it, and it is to them, not buildings, that we must fit our plan' (p.140). Planning from a humanistic perspective suggests 'an avoidance of abstraction in the construction of planning theory, a commitment to experimentation as a method of planning research, and recognition of relationality in planning practice' (Lake, 2017, p.302). The typical and general humanistic approach to planning theory prompts the switch from theorizing participatory planning to the 'inclusive and participatory construction of planning theory' (ibid). It also promotes an important transformation in the practice of planning research. The situated and contextual commitment of humanism is not the 'antibook of objective science but rather the re-visioning of science as an inclusionary experimental attitude. ... [It is] transformed from the unattainable impartiality of the distanced observer to the construction of reality through the inclusion of multiple perspectives' (Harding, 2015, pp. 34–35).

Humanistic concerns have inspired not just a series of critiques against the predominant top-down planning approach, but also a variety of alternative approaches to academic research as well as project practices—these share the trend to 'foreground human experience' (Good, 2017, p.291) and to be treated as the 'source of knowledge' (ibid.) and basis for action (Good, 2017, p.291). Linking planning knowledge to action indeed requires the production of knowledge as an object that can be applied to inform action through planning. As shown in the last chapter, the main research aim is to investigate the concrete meaning and feasibility of people-oriented concept into Chinese eco-city planning practices, thus, an initial question should be addressed here—if yes, and to what extent? Whether the conceptual-practical gap exists in Chinese eco-city projects? From this perspective, this research tends to be the normative slant—it aims to compare the practical level of Chinese eco-city project with the humanistic principle in planning documents, policies and actual construction situations. In this research, the literature review plans to make an argument about why humanistic planning concept should be introduced in Chinese eco-city's project, and why it would be the theoretical basis for the people-place interactive relationship.

## 2.2 Overall Explanation about the Conceptual Framework

The conceptual model interprets the concept of *humanistic planning under the context of ecology* from both micro and macro perspectives to respond to how Chinese eco-city's development consider people-oriented planning concept. Detailed explanation has been illustrated as follows.

This research underscores the holistic approach by recognizing people's diverse needs are required to be considered within broader urban scales, that is, integrating both micro and macro perspectives to understand urbanization development. Specially, it regards urbanization as the macro phenomenon, being comprehended by scrutinizing micro phenomena that constitute it. It believes to promote the process of green urbanization with precision is to ensure people's micro-level needs be identified and be met, such as recognizing the variability of needs among different age groups towards social infrastructures. And this research regards eco-city development as a multifaceted phenomenon that intertwines with ecological civilization. Such development pattern aims to harmonize urban development and economic growth along with environmental conservation. This interplay underscores the imperative to adopt holistic strategies that prioritize the ecological integrity of green spaces while safeguarding people's well-being.

In the second half of this chapter, it essentially discusses linkages among three aspects—urbanization, ecological civilization, as well as people's needs and their behaviours. Referring to the corresponding theory, *Garden City* presents a vision of sustainable development path as the effective response to tackle rapid urbanization challenges and to realize the cognition of basic interaction between people and place. *Ecological Urbanism* serves as a tangible manifestation of principles and also ethos inherent to ecological civilization, embodying a paradigm shift in urban development towards sustainable and humanistic aspects, thereby exemplifying a transformative approach that aligns urban development pattern with the overarching tenets of ecological civilization. *Environmental Psychology* is dedicated to the understanding about the interplay between the physical environment and people's needs, elucidating the ways in which design and configuration of public spaces influence people's well-being.

## 2.3 The Theoretical Intersection of Urban Planning Disciplines: Situating Ecological Urbanism within the Garden City Concept

*Ecological Urbanism* and *Garden City*, these two guiding theories are seen as the complementary approaches to eco-city's planning, by creating communities that are designed to be sustainable and resilient as well as by providing green urban spaces for residents to enjoy and relax. Both concepts contribute to informing decisions for eco-city's development, depending on specific planning goals and the context of a given planning project.

*Ecological Urbanism* is a relatively new approach to urban planning which intends to create sustainable and resilient environment that integrate natural system and urban development. This approach seeks to promote sustainable planning practices and aims to incorporate natural environment into urban design. Key principles of *Ecological Urbanism* theory include promoting resource efficiency, reducing carbon emission, enhancing biodiversity and improving people's health and well-being. While *Garden City* is the classical approach to urban planning, the garden city model seeks to create self-contained communities with a balance of commercial, residential and green spaces, and also with the goal about promoting the healthy and high-quality lifestyle for residents. Furthermore, *Ecological Urbanism* more focuses on advocating urban sustainability and resilience in the face of environmental challenges, with a big-picture development version such as climate change and low-carbon development. While *Garden City* emphasizes on promoting the healthy and high-quality lifestyle for residents, with a more focused angle of view that cares about people's daily lives.

Although two theories still have some key differences, from the perspective of urban development goals, both *Ecological Urbanism* and *Garden City* share to the goal about creating sustainable and liveable cities to create an urban environment that is suitable for people's living. Firstly, both planning principles encompass environmental, social and economic dimensions, thus, by combining these two theories, eco-city planning could be operated more effectively to better address multiple sustainable goals simultaneously; for instance, prioritizing natural resources conservation, promoting social equity and supporting economic prosperity through green industries. Secondly, *Garden City* emphasizes on the incorporation of green spaces, such as parks or gardens, while *Ecological Urbanism* more emphasizes on green infrastructures, which involves integrating natural elements into the built environment. Combination of these two theories has been exactly in line with current development goal for Chinese eco-city's development, to implement green urbanization and digital technology reform. Thirdly, these two concepts all attach importance to people by fostering the sense of ownership and social cohesion—*Garden City* concept advocates community engagement matters in planning decision-making process, *Ecological Urbanism* recognizes that people's needs matter in planning projects.

Overall, suiting *Ecological Urbanism* within *Garden City* to guide eco-city's planning could benefit from a comprehensive framework that addresses humanistic issues, and could enable the creation of cities that prioritize sustainability, resilience and liveability while harmonizing with the natural environment. Moreover, the integration of these two concepts promotes holistic approaches to urban planning, making eco-city vibrant and inclusive. Specific interpretations about how these two theories guide the concept of *humanistic planning under the idea of ecology* are discussed in the following sections.

## 2.4 Ecological Urbanism: Setting out Guidelines to Identify Design Elements Related to People

'Much of the understanding of eco-city and practical initiations of this concept have been influenced by the research of urban ecology' (Yu, 2014, p.103), which studies the interaction between human culture and natural environment and recognizes humans cannot be excluded from urban ecology. As Douglas et al. (2012) stated, 'urban ecosystems are the hybrid of natural and man-made elements whose interactions are affected not only by the natural environment, but also by human culture, personal behaviour, politics and economics' (p.22). Since the science of ecology has developed over time, its emphasis focuses on 'ecosystems of nature interrupted or diverted by the systems of humans' (Hagan, 2015, p.22). *Ecological Urbanism* intends to make this emphasis as far as possible—it overlaps *Urban Ecology* when it considers ecosystem as a syncretic element of human and natural. The focus on *Ecological Urbanism*, however, is on the urban aspect, rather than the ecosystem. At the conceptual level, *Ecological Urbanism* is 'a cultural practice that demands from the built environment, giving a change in thinking' (Hagan, 2015, p.23). On the one hand, *Ecological Urbanism* is a late entry into the line of ecological subfields—it was famous when the book by that title published in 2010 (Mostafavi & Doherty, 2010) as the culmination of a *Harvard School of Design Project*. On the other hand, it has the potential opportunity to build the bridge between urbanism and ecology, one that 'projects and defends design as a vital element in the necessary transformations of cities' (Hagan, 2015, p.5). As Mostafavi and Doherty (2010) define the territory and significance of *Ecological Urbanism* as an urbanism that requires a new sensibility and new opportunities to define a new approach to tackle issues confronting cities and regions—it needs to view 'the capacity to incorporate and accommodate the inherent conflictual conditions between ecology and urbanism' (p.17).

Over the last decades, a series of planning approaches have been adopted to promote urban methodologies, for the purpose of satisfying economic, ecological and social aspects of sustainability. In this research, *Ecological Urbanism* is seen as an innovative promotion to planning approaches, which has been defined as ecological design innovate under the trend of sustainable development. However, this is not to imply that *Ecological Urbanism* is a brand-new theoretical perspective that can help to enhance approaches to urban development. As Mostafavi (2010) stated, '*Ecological Urbanism* is not a totally new and singular mode of design practice, rather, it utilizes a multiplicity of old and new methods, tools and techniques in a cross disciplinary and collaborative approach towards urbanism developed through lens of ecology' (p.13). In other words, even though Mostafavi does not promote innovations in the knowledge of sustainable cities, he focuses on renovating the debate about sustainable development, such as what exactly does it mean to be green or be ecological (Turnbull, 2010).

As a new practical method leading the green urbanization development, Mostafavi does not give a specific definition of urban approach or principles' indicator list for innovative urban models, but he realizes a compendium of brief contributions to renovate attentions about environmental issues, such as population growth or negative impacts on urbanization. He advocates the role of urban design as solutions to seek for a balance between people and nature. Moreover, Mostafavi considers to reduce the effect of current sustainable planning practices on the global scale (Gattegno, 2011), and establishes global and local imperative grounded in design. As he (2010) presented, it remains the problem that 'the moral imperative of sustainability and by implication sustainable design tends to supplant disciplinary contribution. Thus, sustainable design is not always seen as representing design excellence or design innovation' (p.13). From the perspective of theoretical construction, the uniqueness of Mostafavi's theory is about highlighting the importance of urban scale in the same formulation of the term *Ecological Urbanism*, where ecology and urbanism are conjoined to provide 'the knowledge, methods and clues of what the urban can be in years to come' (Mostafavi, 2010, p.13). As Gategno (2011) presented, ecological design does not simply acknowledge 'the fragility of our surrounding ecosystem and the resource limitation, they consider this contemporary state of affairs as the necessary basis for creative design speculation' (p.74).

Arguing for the need of new planning approaches to urbanism to incorporate more ecologically responsible forms of living, Beatley (2000) views a city exemplifying green urbanization as one that 'facilitates more sustainable lifestyles and emphasizes a high quality of neighbourhood and community life' (p.8). As the development of multi-dimensional sustainable human settlement within harmonious and balanced built environment, *Ecological Urbanism* focuses on developing urban environment based on principles of ecological sustainability to promote the process of green urbanization. It draws from ecology to inspire an urbanism that is more sensitive to the environment and more socially inclusive to people, predicated on the assumption that ecology is concerned with the relationship between people and place. This research regards *Ecological Urbanism* as a holistic planning approach, which needs to integrate complex systems and social processes together, and needs to produce combined effects greater than the sum of their separate effects.

### 2.4.1 Environmental Sustainability and Ecological Urbanism

A city, greater than the sum of its parts, should be more comprehensively considered in a collective way. Rather than a piecemeal approach, the holistic and system-based method can provide a more efficient technique to tackle wicked problems. Comparing to the medicine, *Ecological Urbanism* cares about the health of the entire city before addressing individual symptoms. As Rogers (1998) explained, 'building a sustainable city requires a holistic discipline of planning that considers all factors which make up physical, social and economic needs of a community and relates them to the greater environment' (p.53). Specifically, what aspect of *Ecological Urbanism* excels significantly beyond other forms of ecological design is the capacity to engage elements of personal connection—a holistic city is capable of connecting people with nature and places.

*Ecological Urbanism*, a means of adaptation, provides a framework for addressing challenges that threaten humanity and aims to fulfil people's needs for a healthy, safe and sustainable environment. Unlike other planning theories, such as *New Urbanism* or *Landscape Urbanism*, *Ecological Urbanism* has a strong emphasis on environmental ethics as it regards the human as 'part of the environment, and adopts a technical approach to designing cities, which can maximally meet the needs of inhabitants' (Spirn, 2014, p.557). As a term, it embraces three distinct branches of environmental thoughts, including 'environmental protection which focuses on conservation of nature and opposition to pollution, ecology which views the human as enmeshed within environmental systems and addresses itself to human-natural interactions, and environmental justice which considers the impact of environmental change on socially disadvantaged groups and analyses the distributional impacts of environmental policy' (Fainstein, 2010, p.300).

*Ecological Urbanism* is regarded as an evolution of *Landscape Urbanism* because of its expanded focus on aesthetic and functional aspects of urban landscape, also due to the integration of ecological principles and sustainable considerations into urban design process. While *Landscape Urbanism* reimagines urban spaces through a lens of landscape architecture, *Ecological Urbanism* takes this concept further by emphasizing the importance of ecological systems and processes in shaping and sustaining urban environment. *Landscape Urbanism*, a theory of urban design, holds the view that cities are made up of interconnected and ecologically rich horizontal field conditions, rather than an arrangement of objects and buildings (Corner et al., 2003; Waldheim, 2006). Mohsen Mostafavi (2010) defined *Ecological Urbanism* as 'both a critique of and a continuation by other terms of the discourse of *Landscape Urbanism*' (p.114). He strongly promotes this concept in order to seek for an approach which can have the capacity to ease the inherent conflictual conditions between ecology and urbanism. Also, he plans to 'multiply the available lines of thought on the contemporary city to include environmental and ecological concepts, while expanding traditional disciplines and professional frameworks for describing those urban conditions' (Waldheim, 2010, p.114).

Although academic attention has been paid to mainly alternative approaches to eco-urban development and low-carbon transition (Pickerill, 2013), research interests on eco-cites and transitions focused on 'ecological modernization and the role of specific techno-environmental solutions to notions of urban and energy crisis' (Caprotti, 2013, p.11). Compared to other urbanism theories, on the theoretical basis of ecological modernization, *Ecological Urbanism* intends to supplement specific and up-to-date discourses to ecological and social conditions in the contemporary society. Even though eco-city's domestic spaces are regarded as part of new 'filtered communities' enabled by city's green regulations, the lived reality about inhabitants of these spaces is no longer as one of enjoying living conditions for ecological modernization (Spaargaren & Mol, 1992), but of concerns about the diffusion and often invisible risks that still exist in the new city (Beck, 1992). As the theory constantly updated, *Ecological Urbanism* offers an opportunity for planning practitioners to rethink the widespread social influences of environmental movements as well as potential long-term consequences for the profound understanding of cities. As Hodson and Marvin (2010) stated, *Ecological Urbanism* is a new-type urbanism that offers both technological solutions and market frameworks to 'overcome what we would have conventionally understood as limits while anticipating a period of climate change and ensuring continued reproduction under a period of resource constraint' (p.209). Moreover, *Ecological Urbanism* provides a collection of theory and practice that attempts to qualify and ameliorate low-carbon design strategies in China. 'Bottom-up solutions, mobilizing user-led innovation and direct community co-investment, illustrate provocative alternatives to conventional practice' (Marinic, 2012, p.106), which has been in line with the *2030 Carbon Peak and 2060 Carbon Neutrality Goal* that China is now promoting to reduce energy consumption and to promote low-carbon development. Specifically, in order to address environmental challenges posed by the rapid urbanization and to promote sustainable development, *Ecological Urbanism* has provided a framework for designing cities that are ecological and liveable, while the *Carbon Peak and Carbon Neutrality Goal* provides a roadmap for achieving these goals at the national level. Regarding planning practices, the Chinese government has gradually implemented various measures on the basis of principles of *Ecological Urbanism*, such as supporting technology to promote green buildings and encouraging the use of renewable energy sources.

As discussed above, *Ecological Urbanism* promotes the humanistic design by integrating principles of sustainability, environmental consciousness and social well-being into urban development. It seeks to create cities that are not only ecologically responsible but also prioritize people's needs and living qualities for their residents. Firstly, *Ecological Urbanism* takes a holistic approach to urban design, considering interconnections between natural environment, built environment and social dynamics. This approach ensures urban spaces are designed with a deep understanding of how they impact people's lives. Secondly, *Ecological Urbanism* and the humanistic design intersect through their shared emphasis on social well-being. By integrating ecological principles with people-oriented planning approaches, cities could become more liveable for people to thrive.

Reasons for advocating *Ecological Urbanism* concept are compelling as one of main research objectives is to focus on future's humanity, that is, whether people can adapt their behaviours to meet current urbanization challenges. For example, in terms of the driving force for eco-city's development, as discussed in the previous chapter, Chinese long-existing urbanization development model is seen as a product of traditional industrial era, while green urbanization is a leap from the traditional state to the new one. This change has given the breeding ground to Chinese rapid eco-city development. Being a brand-new conception proposed in the year of 2010, the *Ecological Urbanism* theory has its own limitation—as Marinic (2012) argued, the book *Ecological Urbanism* edited by Mohsen Mostafavi operates more like 'a decade-long encyclopaedic compilation of trans-disciplinary environmental thought' (p.105). It is closely associated with international planning projects, which means exploring in real-world practices to some extent lacks of solid theoretical foundation. Thus, evolution and development of this theory should be tested through time in order to see whether *Ecological Urbanism* contributes to the contemporary eco-city planning in the face of multiple eco-emergencies. As is known, designers have essential roles, not merely in designing safer and healthier urban habitats, but in making the entire environment legible and tangible.

### 2.4.2 Planning for People

*Ecological Urbanism* has presented the comprehensive global range of urban interventions, strategies and proposals that frame a radically interdisciplinary vision of future sustainability. It grows out of ideas and actions in landscape, architecture, urban design and urban planning. This concept has a long history as it needs a foundation of knowledge to support its theoretical content and to build models for demonstrating its benefits. This research selects some planners, even though they are all pretty much early to the mid-twentieth century, situating their work within this research because they contribute to the humanistic planning concept by prioritizing people's needs and social well-being. Moreover, their ideas have influenced planning practices that value sustainable design and have enhanced the connection between people and place. In short, Patrick Geddes believes a holistic approach to planning should integrate social, cultural and environmental factors. Lewis Mumford advocates human-scale cities that prioritize the quality of life for people. *The Image of the City* proposed by Kevin Lynch focuses on how people perceive and navigate urban environment—by understanding how individuals mentally map their surroundings, planners can create more people-centred and navigable cities. Ian McHarg's design with nature approach emphasizes the integration of ecological principles into urban planning. He believes planning should be based in a deep understanding of natural systems and their interactions with people's activities. Jane Jacobs stresses the importance of preserving social fabric of neighbourhood and allowing them to evolve organically.

Lewis Mumford advocates new urban form must 'include the form-shaping contributions of nature, as well as those of human history and culture, with the complex interplay of groups, corporations, organizations, institutions, personalities' (Mumford, 1968, p.164). Mumford influenced Kevin Lynch and Ian McHarg—for Lynch, the city is first and foremost a human habitat, and he judged the good city form by how well it sustains human lives. Lynch stressed the significance of how people perceive the city, namely the human perception to understand the sense of place. He explored the role which natural features play in enhancing the 'identity, legibility, coherence, and immediacy of urban form' (Lynch, 1981, p.11). Moreover, Lynch suggested criteria of the good city form—how well urban form sustains life (vitality), how clearly it is perceived in space and time (sense), how well environment and behaviour fit, and whether these standards are presented in a manner that provides 'access, control, efficiency and justice' (ibid.). For McHarg (1967), like Geddes, he believed 'any place can only be understood through its physical evolution' (p.105). As a prerequisite for urban planning and design, he proposed the ecological inventory, checklist for the interrelated systems, being seen as a 'diagnostic tool with which to identify problems and opportunities that might otherwise missed' (Spirn, 2000, p.108). From this point of view, for McHarg, design is an evolutionary strategy, namely a means of adaptation. Another contributor to *Ecological Urbanism* is Jane Jacobs, author of the classic book *The Death and Life of Great American Cities*. Like Lynch, she focused on the city as a human habitat and regarded urban design as a way to support and fulfil people's basic living needs. Jacobs' work mainly focused on analysing the influences of new modernist development in cities like New York, rather than newly built eco-cities. However, newly built eco-cities are often planned along modernist lines, in terms of their planning layout and architecture. To seek common points while reserving difference, due to such parallel, this research still adopts Jacobs' work as her insightful critique about the lack of attention paid to social sustainability has concentrated on analysing the dehumanizing impacts.

### 2.4.3 Suitability as the Humanistic Planning Theory

*Ecological Urbanism*, combining and intensifying efforts of other valuable sustainable movements and disciplines, has presented a viable alternative to conventional planning theories. As demonstrated, its content is expressive of a harmonious society, yet it is not 'overly fanciful, it does not paint the picture of an unrealistic eco-topia' (Griffith, 2013, p.34). It has reflected the worldwide ideological trend, which intends to make a contribution to current urban planning and design. Attentions based on the global climate change have been embodied, and the academic has created subversive reflections on contemporary urban issues, especially the unprecedented scale of urbanization and corresponding environmental pressures. Even though explorations of ecological cities have a long history, different kinds of concepts and practices (e.g. *New Urbanism*, *Landscape Urbanism*, *Liveable City*, *Compact City*) all aim to tackle for negative impacts on modern cities. However, what is unique for *Ecological Urbanism* to be distinguished from others is it regards the city as an urban ecosystem, trying to create the harmonious, efficient and sustainable human habitat with respect to social, economic, cultural and technological aspects. From this point of view, considering people's needs, the main objective of *Ecological Urbanism*, is closely associated with the people-oriented planning concept. Moreover, its systematic approaches and multi-perspective practical explorations indicate that this concept does not define a model or a development route for future's ideal cities, nor provides a manual for eco-city construction. Indeed, it aims to offer various designing methods and technical references to realize development goal about ecological civilization (Yu, 2010, p.2).

As discussed above, *Ecological Urbanism* is an approach to urban planning and design that emphasizes the integration of natural systems into urban planning process. *Garden City*, on the other hand, is an urban planning concept that originated in the early twentieth century and emphasizes the importance of incorporating green spaces into urban design. The garden city model seeks to create self-contained communities with a balance of residential, commercial and green spaces, with development goal for promoting healthy and high-quality lifestyle for urban residents. Referring to their relationship, *Ecological Urbanism* and *Garden City* are seen as complementary approaches to urban planning and design. For instance, the eco-city development could incorporate these principles by creating self-contained communities that are designed to be sustainable and resilient, while also providing green spaces for people to enjoy. By combining these approaches, the eco-city might create liveable, sustainable and healthy environment that meet residents' living needs.

## 2.5 Garden City: The Theoretical Foundation of eco-City

### *2.5.1 Main Concepts of the Theory*

The concept of *Garden City* has been deemed as being born with Ebenezer Howard even though various 'villages of vision' (Darley, 1975) for spatial planning boomed since the nineteenth century. Rosenau (2007), for instance, situated Howard within the *Ideal City* tradition, placing him among 'utopians of the 19$^{th}$ century and prefiguring the futurist of the early twentieth century' (p.36). His book *Garden Cities of To-morrow*, published in 1902, was a powerful response to severe overcrowding issues and diseases in large cities, especially London at the time of specific social background, where high population densities as well as poor sanitary conditions led to miserable lives and early death. After publishing, it soon became a 'classic reference work' (Joss et al., 2015, p.81) and profoundly inspired planning theories and policies in twentieth century worldwide. From this point of view, core values of *Garden City* theory are regarded not only as the conceptual and practical proposal for tackling with negative effects of the unprecedented urbanization, especially during the industrial revolution period, but also a political work in some ways as it is a critique about the ills of industrial urban society. As Parsons and Schuyler (2002) stated, Howard attempted to achieve things that other planning proposals failed to do, that is, 'a workable book between the visionary and the practical, and one grounded in a pragmatic but highly communitarian economic structure' (p.6).

To achieve the harmonious spatiality, Howard first proposed the *Garden City* concept in his influential book *Tomorrow: A Peaceful Path to Real Reform*, stating 'all advantages of the most energetic town life, with all the beauty and delight of the country, may be secured in perfect combination' (Howard, 1902, p.46). His intention was to be both conceptual and practical, leading him to propose a concrete model of garden city, outlining urban infrastructures. Moreover, the core content of Howard's conception is seen as a 'joyous union' of town and country, which is deployed by his 'visual aid' called *Three Magnets Diagram* (See Fig. 2.1). Howard used this simple analogy of the magnet, capable in 'equal measure of attracting or repelling a small iron bar, representing the people' (Ward, 2016, p.16). He revealed both desirable and undesirable features in this town and country magnets, analogous to both positive and negative poles of a magnet. He then proposed his town-country magnet which would combine those positive features compared to other twos. Specifically, values expressed in the town-country magnet have indicated various social concerns, such as high amenity living environment and the access to nature.

Under the content of a combination of town and country, Howard proposed the creation of new settlement pattern—freestanding garden cities would ultimately replace large cities. Although Howard did not describe any design details, this target plan still clearly indicated he planned his garden city to be an integral part of a wider urban region instead of being 'entirely self-contained and inward-looking' (Ward, 2016, p.14). Specifically, within each garden city, people could live, work, shop and enjoy different range of leisure activities as well as public and private services they needed. Urban growth would be strictly controlled

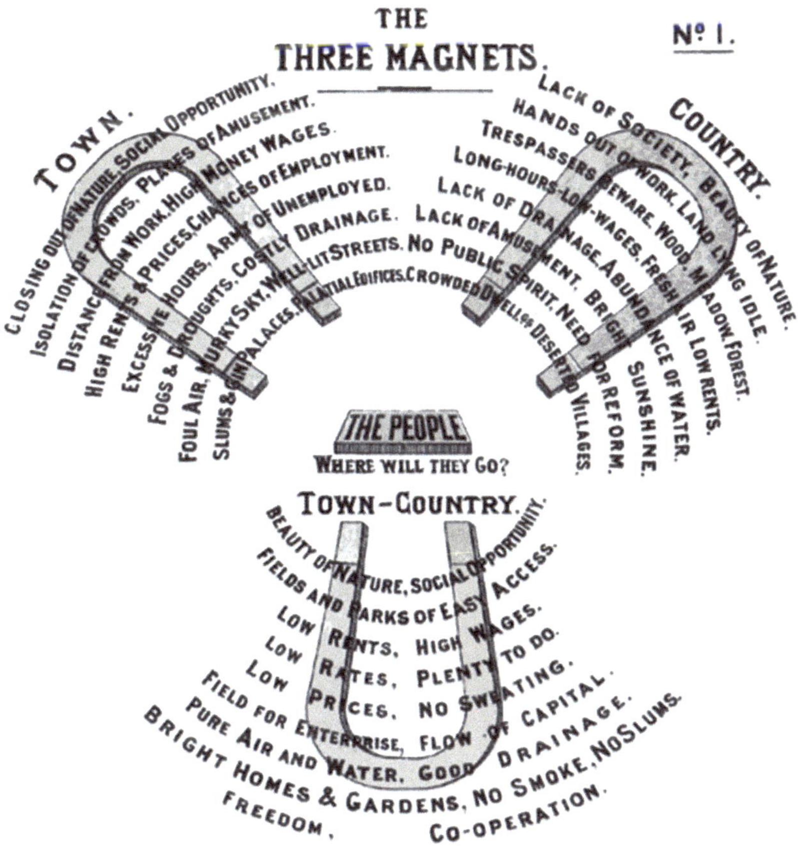


**Fig. 2.1** Three Magnets Diagram proposed by Howard (*Source* Figure is from Spatial Agency Web)

because an upper limit on the number of inhabitants has been set for each garden city, and around each one there would be a well-protected rural belt. Moreover, in order to keep with this 'joyous union', both economic and social separation for urban and rural dwellers would also be minimized and would achieve the win-win outcomes. In the proximity to the town, rural dwellers could make use of social facilities and services, meanwhile, town dwellers could easily get access the natural scenery of countryside if the nature of town itself was insufficient. As Howard advocates for the social justice, designing garden cities should be socially equal, with different classes of people as citizens of the same garden city and accessing to social resources.

### 2.5.2 *The Cognition of Basic Interaction between People and Place*

In terms of urban design, although Howard was certainly influenced by utopian conceptions, his outlook version was 'firmly rooted in the real-world situations, and also in the urgent human needs for an alternative to the industrial cities of late Victorian Britain' (Hugel, 2017, p.2). Specifically, under the special history background, the *Garden City* concept to a large extent was a response to the overcrowding, disease and general misery he experienced in London where high population densities, poor housing and sanitary conditions caused miserable lives and early deaths. As Parham (2013) evaluated, Howard took the radical step to propose a settlement form that offers the self-sustaining community for garden city's residents (p.2). From this point of view, the original planning intention for *Garden City* concept can be interpreted as seeking for better city and better life—caring about people's basic needs and improving their living standards to achieve more liveable human habitant.

Regarding the relationship between people and place, core planning objective of the *Garden City* movement is to create places in which people can have a genuine stake in the development process and are able to live healthy social lives. Moreover, it believes where people come forward with a collaborative and co-operative approach, the result to a large extent will be high-quality and well-designed places. The most powerful evidence to prove this is the core of three magnets diagram—Howard asked, where will the people go? It is of great importance to rethink the *Garden City* concept for the twenty-first century, and in particular to rediscover its benefits about delivering attractive, liveable and sustainable new communities—'the breakthrough and more comprehensive response to the grim, environmentally degraded industrial cities symbolically associated with the grid forms' (Dobbins, 2009, p.40). That is, Howard's transcendent aim was to create an urban settlement that would be better for everyone living there, not just the landowner or the developer. His garden city proposals in 'diagrammatic abstractions projected a new way for conceptualizing cities, which picked up from utopian notions of embodying physical and social content as the drivers for new forms' (ibid).

### 2.5.3 Suitability as the Humanistic Planning Theory

Howard's idea reflects an awareness of fundamental interactions between people and their physical environment, as diagrams, he attempts to rationalize these interactions into formalistic geometric patterns (Dobbins, 2009, p.4). These ideas were first found to form in interpretation by Raymond Unwin and Barry Baker in Letchworth, and later the New Town movement boomed. Moreover, even though as an adherent of *Garden City* theory, Patrick Geddes purposed a different perspective compared to Howard's ideas after absorbing key points. That is, Geddes had been advocating the importance of understanding and respecting what was already there, both physical and cultural, in order to deal with urban design problems in his planning practices. As Dobbins (2009) stated, '[Geddes] is a purer advocate for the organic, indigenous, incremental approach' (p.42), including the vital relationship between people and place. In a sense, comprehensive and inclusive theoretical foundations advocated by these urban planning pioneers during the twentieth century, had and still have significant impacts on designing human settlement pattern, and also on enhancing the sustainability of spatial quality.

The origin of *Garden City* theory, illustrating the complex interplay of spatial integration, has shaped modern urban planning for more than a century. It aims to invent a new type of human settlement where both scale and planned order can promote the development of rational communities. Basic planning principle conveyed by Howard that people need to live in better places, not just better homes, remains powerful today. It cannot be claimed that Howard's unique proposal of *Garden City* theory is suitable in its wholeness, however, fundamental characters such as human-scale settlement and people place relationship, have kept a continuing attraction. To some extent, the suitability of *Garden City* idea for contemporary urban planning discipline is much significant than it was roughly a century ago as current attention has become much more focused on the sustainable development and environmental ethics, which is comparatively similar to Howard's ideas about harmony and balanced social city. As Hardy (1989) demonstrated its city version as 'settlement of a manageable size with a sense of identity, the provision of a good living environment' (p.188) to secure benefits and remain attractive aspirations for the whole community. Furthermore, former *Garden City* theory advocates, especially Lewis Mumford, proposed the thought about human-centred living environment to be affordable, attractive and accessible to dwellers, which still remains as the priority in eco-city initiates today.

### 2.5.4 *The Continuation of Garden City Theory*

As illustrated above, *Garden City* theory is the theoretical foundation of eco-city, and constructing modern eco-city is the continuation of the garden city development. So, the comparative analysis between *Nine Garden City Principles* (See Table 2.1) listed by Town and Country Planning Association and *Ten Eco-City Principles* proposed by Urban Ecology (See Table 2.2) has been applied to further prove the relationship between these two concepts. In his article *Dimensions of the Eco-City*, Mark Roseland (1997) listed ten principles to create eco-city, defining the notion as 'the most durable kind of settlement that humans are capable of building and a city that provides an acceptable standard of living without depleting the eco-system' (Lye and Chen, 2010, p.3). The aim of these ten principles has been summarize as enabling residents to live a good quality of life while using minimal natural resources. Referring to the *Garden City Principles* listed by the Town and Country Planning Association, they form an indivisible and interlocking framework for the delivery of high-quality places. Moreover, these principles are seen as the distillation of key elements that have made the construction of garden city be more suitable for the twenty-first century development situations. The aim of these nine principles is seen as to construct a holistically planned new settlement which 'enhances the natural environment and offers high-quality affordable housing and locally accessible work in beautiful, healthy and sociable communities' (Town and Country Planning Association Website, no date shown).

How to plan and build our cities to be more sustainable and liveable without damaging ecological surroundings also have been a focus to planning practitioners for more than a century. Since the 1990s, different planning theories have emerged (See Fig. 2.2), and in the first half of the twentieth century, urban planning was influenced by utopian ideas. Ebenezer Howard has been widely known as the person who starts the new wave of utopian thinking, and his utopian vision of a planned community has been the focus of attention ever since. What is more, he aims to invent a brand-new type of human settlement whose size, scale and planned order would contribute to the development of rational

**Table 2.1** Representative Work of Humanistic Urban Design

| *1960s* | *1970s* | *1980s* |
|---|---|---|
| 1960, Kevin Lynch *the image of the City* | 1971, Jan Gehl *life between buildings* | 1980, William H. Whyte *the social life of small urban spaces* |
| 1961, Jane Jacobs *the death and life of great American cities* | 1977, Christopher Alexander *A pattern language* | 1987, Christopher Alexander. et al. *A new theory of Urban Design* |

(*Source* Information is from Matan and Newman, 2016, p.30)

**Table 2.2** Main Planning Ideas of Ecological Urbanism

| *Planners* | *Main Planning Ideas* |
|---|---|
| Patrick Geddes | Each city and its surrounding countryside as an 'evolving organic whole' whose future plan should be based on an understanding of its natural and cultural history. |
| Lewis Mumford | New urban form must include the form-shaping contributions of nature, as well as those of human history and culture, with the complex interplay of groups, corporations, organizations. |
| Kelvin Lynch | Suggesting criteria of the good city form—how well urban form sustains life (vitality), how clearly it is perceived in space and time (sense), how well environment and behaviours fit. |
| Ian McHarg | Design is an evolutionary strategy, namely a means of adaptation. |
| Jane Jacobs | City as a human habitat and regarded urban design as a way to fulfil people's needs. |

(*Source* Information is summarized by the researcher)

communities, not only for an industrial age but also for the contemporary society. It can be said the enduring power of *Garden City* concept is timeless ideals, which are also ideals for modern sustainable development and eco-cities to be more humanistic. It is worth mentioning that *Ecological Urbanism* is not listed in the table not only because it is not a specific theory, but also due to its limitation as discussed above—even though being a new planning conception proposed in the year of 2010, it is closely associated with some international planning projects, which means exploring in real-world practices to some extent lacks of solid theoretical foundation. Thus, the evolution and development should be tested through time to see whether *Ecological Urbanism* theory contributes

**Table 2.3** Ten eco-city principles (urban ecology)

| |
|---|
| Revise land-use priorities to create compact, diverse, green, safe, pleasant and vital mixed-use communities near transit nodes and other transportation facilities. |
| Revise transportation priorities to favour foot, bicycle, cart, and transit over autos, and to emphasize 'access by proximity. |
| Restore damaged urban environments, especially creeks, shore lines, ridgelines and wetlands. |
| Create decent, affordable, safe, convenient, and racially and economically mixed housing. |
| Nurture social justice and create improved opportunities for women, people of colour and the disabled. |
| Support local agriculture, urban greening projects and community gardening. |
| Promote recycling, innovative appropriate technology, and resource conservation while reducing pollution and hazardous wastes. |
| Work with businesses to support ecologically sound economic activity while discouraging pollution, waste, and the use and production of hazardous materials. |
| Promote voluntary simplicity and discourage excessive consumption of material goods. |
| Increase awareness of the local environment and bioregion through activist and educational projects that increase public awareness of ecological sustainability issues. |

(*Source* Information is from Roseland. 1997. pp. 197–198)

to the contemporary eco-city planning practices in the face of multiple eco-emergencies.

Chinese eco-city programmes have drawn inspiration from the *Garden City* concept, by promoting sustainable urbanization and improving residents' life quality. One example of the *Garden City* concept applied into Chinese urban planning is *Shanghai One City Nine Towns Project* launched in the early 2000s. It is an economic development plan proposed by the *Shanghai Planning Commission* in the year of 2001, and urban planning was essentially completed in 2007. Echoing the idea of *Garden City* theory, planning purpose is to relieve the overcrowding of overdeveloped city centre and to prevent urban sprawl around Shanghai. This project has intended to provide a combination of housing, education and commercial investment opportunities in satellite new towns built on the farmland outside the city. While the *Garden City* concept has had some influences on Chinese urban planning, it is important to note there are unique challenges to apply this concept in China. For instance, under the traditional urbanization development mode, high population density but limited land space makes it difficult to create self-contained communities

**Table 2.4** Nine garden city principles

| |
|---|
| Land value capture for the benefit of the community. |
| Strong vision, leadership and community engagement. |
| Community ownership of land and long-term stewardship of assets. |
| Mixed-tenure homes and housing types that are genuinely affordable. |
| A wide range of local jobs in the Garden City within easy commuting distance of homes. |
| Beautifully and imaginatively designed homes with gardens, combining the best of town and country to create healthy communities, and including opportunities to grow food. |
| Development that enhances the natural environment, providing a comprehensive green infrastructure network and net biodiversity gains, and that uses zero-carbon and energy-positive technology to ensure climate resilience. |
| Strong cultural, recreational and shopping facilities in walkable, vibrant, sociable neighbourhoods. |
| Integrated and accessible transport systems, with walking, cycling and public transport designed to be the most attractive forms of local transport. |

(*Source* Information is from TCPA Website)

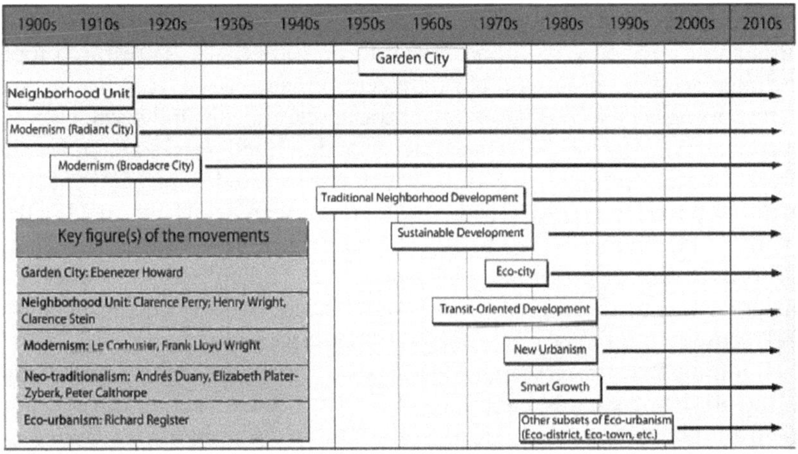


**Fig. 2.2** Theories emerged since early twentieth century (*Source* Figure is from Sharifi, 2016, p. 3)

surrounded by green public spaces. Additionally, Chinese rapid urbanization process has led to the displacement of rural residents and the loss of farmland, which can some have significant social and environment impacts.

## 2.6 Environmental Psychology: To Understand People's Needs and Behaviours

*Environmental Psychology*, a 'pre-paradigmatic stage' (Lang et al., 1974, p.11), refers to 'the relationship between people and the physical environment' (Bell et al., 1984, p.5). It is regarded as the study of transaction between people and physical settings, looking exclusively at the research-based information in terms of complex interactions between environmental factors and people's feeling or actions. In this transaction, individuals can change the environment, their experience and behaviours are influenced by the environment. It includes both theoretical and practical researches, which attempt to make the built environment more humane and to improve the people-place relationship to be more sustainable and more harmonious (Gifford, 2007, p.4). Its research topic includes studies about people's perception and cognition, social and anthropological psychology, study of social relationship and social cultures (Lang, 1987, p.21). Moreover, he also described this concept as the psychological study of behaviours, which is associated to physical environment of people's everyday lives. Academic researches (Lang, 1987; Bell et al., 1984; Bonnes & Secchiaroli, 1995; Bechtel & Churchaman, 2002; Gifford, 2002) of environmental psychology highlighted this concept is included within the topic of behavioural science—how people perceive, feel, sense and interact with their surrounding environments. It is worth mentioning that this research regards *Environmental Psychology* as being in the pre-being paradigmatic stage due to its position within the development of disciplinary integration. In the case of *Environmental Psychology*, being in a pre-paradigmatic stage suggests the field is still developing that defines its core concepts and principles into integrating with urban planning discipline. In recent decades, researchers in related fields are likely exploring different theories and methodologies to better understand complex interactions between individuals and their environment. Over time, with more research being conducted, consensus starts to form around certain theories and approaches, the field may transit into a more mature and paradigm-driven stage for disciplinary integration.

*Environmental Psychology* studies the effect of social environment and the quality of interpersonal relationship we share on human behaviours and interactions. It also studies different approaches to enhancing people's pro-environmental behaviours by their habits or attitudes, in order to minimize negative effects about people's actions on the environment. At the initial development period during the 1980s, the majority of environmental psychologists focused their research on the internal mechanism within people's psychology rather than people-environment transactions. However, considering enormous investment in both construction and maintenance of the physical environment as well as enormous cost of misusing natural resources, the long delay before people-environment relations received adequate attention. Since the late 1960s, research work was stimulated by the recognition of environmental problems, such as pollution, energy shortage, unsuitable buildings. Other research motivated by 'pure curiosity about how and why people act and feel in their built and natural settings' (Gifford, 2007. p.1). Thus, to understand people-environment transactions and to use this disciplinary knowledge for solving a wide variety of real-world issues, *Environmental Psychology* may seem to be indistinguishable from the main core of psychology since most psychologists focus their research on the relationship between environmental stimuli and human responses. However, what make environment psychology apart are its goals and principles shown as follows. '[Firstly], improving the built environment and our stewardship of natural resource. [Secondly], studying everyday setting or close simulations of them. [Thirdly], considering the person and the setting to be a holistic entity. [Fourthly], recognizing individuals actively cope with and shape the environment; they do not passively respond to environmental forces. [Fifthly], working in conjunction with other disciplines' (Gifford, 2007. p.4).

*Environmental Psychology* mainly focuses on people's psychology and their surrounding environment, and this theory believes different scales of the built environment has different impacts on human behaviours. As people's behaviour include both physical and psychological aspects, it is not only dominated by physical needs, but also influenced by psychological needs, and the latter is more complicated. For example, same built environment can still have different impacts on different groups of people regarding their gender and age. Thus, in the face of people-place interactive relationship characterized by its complex diversity, knowledge about ecological environment is vital to studying relationship between physical environment and people's behaviours (Barker, 1968). Research about environmental planning combined with the behavioural design helps to understand the people-place interaction, and also helps to promote a more people-oriented planning approach. In this research, it is of great importance to understand and to clarify such relationship. 'The ecological approach has opened a new door for looking at the relationship between individuals and the environment' (Cosco, 2006, p.17). The most crucial feature of such ecological approach is the affordance concept (Gibson, 1979) that highlights the reciprocal relationship between people and the environment. Closely linked to the concept of affordance advocated by Gibson is the theory of behaviour setting firstly proposed by Barker (1976).

Affordance theory proposed by Gibson in 1986 has been seen as a pioneering ecological thought, aiming to solve people-place coordinational issues from the microscopic psychological mechanism, and to construct the embodied people-place interaction from the perceptual level. Gibson proposed this concept by using the framework of *ecological perception* to define the purpose of perception as perceiving affordances, as he (1986) believes 'seeing something is to see what could be done with that thing' (p.12). From this perspective, the affordance that environment provides has been regarded as the precondition for activities. The concept of affordance focuses on describing the environment from people's behavioural aspect, by providing theoretical support to analyse residents' outdoor activities in public spaces. Also, it is of great importance to understand the environment-behaviour interaction as it helps to explore physical environment's potential impacts on people's behaviours, as well as to understand environmental attributes that are closely associated with people's specific behavioural responses (Gibson & Pick, 2000).

Gibson (1986) claimed the affordance concept was inspired by the thought of Gestalt psychologists, affordances do not change even though people's needs change (p.139), as they are not properties of phenomenological world that depend on the state of people, rather, they are ecological phenomenon that indeed exist in the environment. As Gibson (1982) emphasized, 'the perception of what something affords should not be confused with the colouring of experience by needs and motives' (p.410). His viewpoint of ecological psychology has been understood as a critique on 'mechanistic metaphors' (Withagen et al., 2012, p.250) that dominate the ideological trend in psychological field since the seventeenth century (Reed, 1996; Withagen and Michaels, 2005). His revolutionary theory of perception predated the concept of affordance. However, the latter has become the central part of his ecological theory in terms of perception. In this research, two points of the affordance theory are worth considering—the environment itself is meaningful and consists of opportunities for people's behaviours. Specifically, ever since mechanization of the worldview in the seventeenth century, the environment is generally conceived as meaningless, consisting merely of matter in motion (Neisser, 1967). However, Gibson argued the environment is meaningful as it consists of affordance. As he advocated, 'meaning or value of a thing consists of what it affords' (Gibson, 1982, p.407). He also linked the affordance theory to practical cases by developing a series of ecological approaches in order to understand how these actional possibilities can effectively be perceived. Moreover, Gilson overturned the idea that the environment 'comprises stimuli that causes the perception and action' (Withagen et al., 2012, p.251). In his framework, the environment has been conceived not only as a collection of causes, but also as a manifold of action possibilities. In other words, actional possibilities do exist owing to the relation between properties of people and the environment, which has given the inspiration about how these two objects can be mutually reinforced and effected in this research.

## 2.7 Environmental Psychology and Urban Planning: Where Can Twain Meet

The use of *Environmental Psychology* theory in urban planning discipline is not too commonplace yet, but as the field grows and matures, more and more psychologists consult on the designing of buildings or spaces for a wide variety of purposes. As Heimsath (1978) depicted the possibility of interdisciplinary trend, it requires 'sophisticated insight that does not come from designers, they came from a new breed of psychologists who look into urban design fields, briefly at first, and then more intently as they discern that we are constructing buildings without knowledge of people' (pp. 40–41).

Over the years, *Environmental Psychology* has become part of an interdisciplinary field, its research focus mainly relates to the environment-behaviour study, which involves individual's experience as one aspect and environmental perception as another one. This applied psychological theory attracts anyone who interests in the interface and the relationship between people's behaviours and the physical environment, which unquestionably includes planners. Regarding Chinese ecological civilization construction, planning is by definition a multi-disciplinary endeavour, which is open to social science and focuses on the interrelationships among different aspects of the environment. This section aims to examine some commonalities and differences between environmental psychology and urban planning.

Using the metaphor put forward by Kidder and Fine (1987), lens applied in urban planning is more like a zooming-out one, while lens applied in environmental psychology is more like a zooming-in one. Three differences between environmental psychology and urban planning are shown as follows. Firstly, the main one is about consideration of the individual—urban planning usually gives priority to the aggregation or collective of people (Jones, 1996), while the environment-behaviour approach mainly emphasizes on different groups of people and the degree to which the environment fits their needs. Even though these two disciplines are basically classified into the humanistic field, environmental psychology more closely relates to people and their needs concretely. Secondly, differences between these two disciplines focus on the macro-micro perspective. The focal point of environmental psychology mainly places particular emphasis on the micro level or on relatively small-scale environment, such as the theory of human development proposed by Bronfenbrenner. As he (1994) demonstrated, 'human development is the product of interaction between the growing human organism and its environment' (p.16), which considers potential influences of all systems that play the role in impacting the lived experiences of individuals no matter how remote the influence itself is. Urban planning mainly focuses on the macro level or on relatively large-scale environment, both in geographic, social and economic terms. In a manner of speaking, ecosystem would be the most obvious linkage between these two disciplines. Thirdly, from the time frame perspective, environmental psychology mainly studies 'the past and the present, and only sometimes is willing to hazard a guess as to what this means for the future, while urban planning deals with the future – its mandate is to focus on that' (Churchman, 2002, p. 6).

Jones (1996) attempted to bridge two disciplines by applying Fishbein and Ajzen's theory of reasoned action to environmental planning issues. She realized the significance of understanding individual's thinking and behaviours for planners, in order to 'recognize linkages between attitudes and perceptions of behavioural control' (Churchman, 2002, pp. 12–13). Nevertheless, the major challenge faced by the environmental psychology is about making its academic research more useable in other fields relating to environmental aspects. As Chokor (1987) criticized, the failure of environmental-behaviour research to affect urban planning or urban design practices was mostly blamed on the 'inability of environmental psychologists to make their research relevant to practising urban designers, relating to real-life design issues rather than artificial laboratory situations' (p.23).

## 2.8 Theoretical Intersection to Support Humanistic Planning Concept under the Context of Ecology

In China, eco-city planning intends to realize the harmony between people and living environment by means of traditional ecological thought; the most representative one is *the unity of man and nature* (天人合一). Gibson's theory of affordance starts from the origin of people's perception and behaviours, trying to reposition the relationship between people and the environment. It also examines essential attributes about ecological design and clarifies the purpose of ecological design. This research advocates people-oriented eco-city planning needs to absorb the essence of Gibson's ecological thought due to two main reasons. On the one hand, environmental design based on the theory of affordance is a mechanism for predicting the ecological relationship between people and place by means of presupposing appropriate natural and social affordances within the context of environment. On the other hand, understanding the essential ecological attribute for environmental design ought to start from the nature of human beings by taking people's behaviours and their interaction with the surrounding environment as the fundamental purpose of eco-city's planning, which is seen as the connotation of *humanistic planning under the ecological concept*.

More specifically, as briefly mentioned, the theoretical framework in China differs from those of the global mainstream theories, represented by American and Northern European perspectives, in their approach to the humanistic planning—in the global mainstream context, the focus of humanistic planning is much on individual's pursuit of self-realization and self-expression, which is based on the idea that individuals have innate capacities for self-realization. Therefore, the primary goal of humanistic planning is mainly to create the social environment that aims to foster individual's growth and development. Maslow's hierarchy of needs is the case in point. It emphasizes the importance of meeting individual's needs as the foundation for people's fulfilment, which is the reason why public participation becomes one of the research focuses for the humanistic planning. Chinese theories, such as Taoism or Confucianism, stress the importance about both balance and harmony between people and nature. The individual is seen as part of social and ecological system to be interconnected with the community welfare as a whole for pursuing the collective well-being.

Regarding the theoretical interpretation discussed above, even though there are some obvious differences between global mainstream theories and the Chinese context, this research attempts to gain a brand-new understanding of *humanistic planning concept under the context of ecology* for the purpose of guiding Chinese eco-city development to be more people-oriented. In this research, the people-place relationship will be identified to further understand the humanistic planning concept in eco-city development. In order to comprehensively analyse this relationship, the conceptual framework will be constructed to guide appropriate research methods for data collection and analysis, also, it will be mainly used to interpret and to organize research findings. This integrated framework might contribute to filling the theoretical-practical gap by understanding in a multi-disciplinary way, and to investigating the concrete meaning and feasibility of global humanistic planning theory into Chinese eco-city planning practices, including its core connotation and implications in a broader sense of *humanistic planning under the context of ecology*. It is worth mentioning that there appears to be a gap in the integration of planning theories (*Ecological Urbanism* and *Garden City*) and applied social psychology (*Environmental Psychology*). Main reason about this kind of gap could be regarded as these theories, individually, could not respond to humanistic planning under the context of ecological development effectively. As *Table 2.8* shows, each theory, individually, has its own imperfection to construct the theoretical framework in this research—*Garden City* lacks the transformation from the theoretical advocacy to design-pattern language, *Ecological Urbanism* lacks the systematic research on the people-place relationship and *Environmental Psychology* lacks the specific ecological research from the perspective of urban planning even though it emphasizes on the relationship between people and place.

As shown in Fig. 2.3, this integrated theoretical framework delves into interconnections among developmental trajectory of urbanization, paradigm of ecological civilization and people's needs. It plans to explicate the concept of humanistic planning within the ecological context, considering both micro and macro perspectives. The role of this framework is to address how eco-city development incorporates people-oriented planning concept, and the integrated theoretical framework itself improves the existing people-oriented planning concept in two ways. Firstly, it examines the concept of humanistic planning not only within urban planning and design area, but also broadly extending to psychological area. That is, as the planning goal of people-oriented eco-city is influenced by two different factors, namely people and the environment, this research suits the people-oriented planning concept under the context of ecology in the broader context with which it interacts. Secondly, this theoretical framework more explicitly builds the dynamic nature of humanistic planning, including people-oriented concept and the interactive people-place relationship, which are two core elements for interpreting the people-oriented planning Fig. 2.4.

**Table 2.5** Foci and gap to construct the theoretical framework

| *Theory* | *Discipline* | *Foci* | *Gap* |
|---|---|---|---|
| Garden City | Urban planning | Provide the coordinated ecological-urbanization contradictory value orientation | Lack of the transformation from theoretical advocacy to design-pattern language |
| Ecological urbanism | Urban Design | Provide the spatial means to meet ecological design goals | Lack of systematic research on the people-place relationship |
| Environmental psychology | Applied social psychology | Analyse the influencing mechanism of people's behaviour in the space | Lack of the specific ecological research from the perspective of urban planning |

**Fig. 2.3** Theoretical framework of humanistic planning (*Source* Chart is drawn by the researcher)

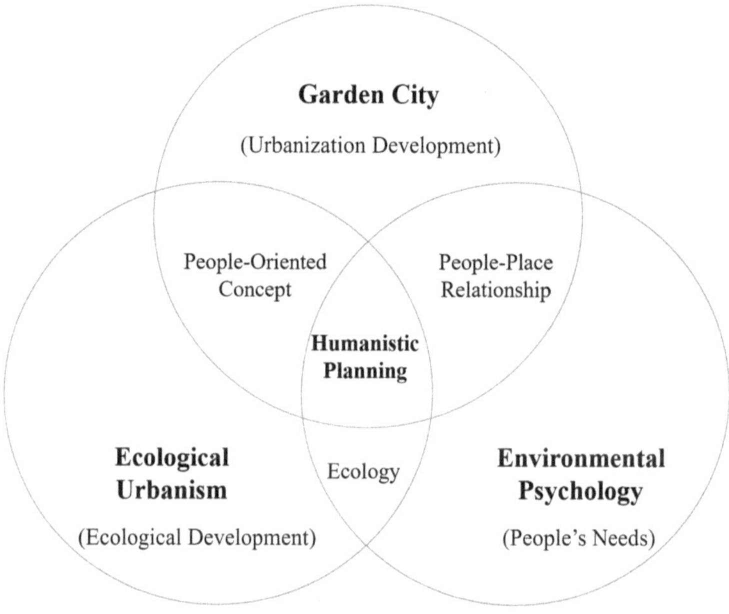


**Fig. 2.4** Concentrations of theory and intersection in-between (*Source* Figures are drawn by the researcher)

## 2.9 Conclusion

This chapter discusses the people-oriented concept by reviewing different theories in three disciplines—urban planning, urban design and applied social psychology. It integrates these three theories into the integrated theoretical framework to better illustrate *the humanistic planning under the context of ecology*.

*Ecological Urbanism* considers the city with multiple instruments and with a worldview that is fluid in scale and disciplinary focus, that is, urban design provides the synthetic key to connect ecology with an urbanism that is not in contradiction with its environment. For instance, *Ecological Urbanism* regards 'urban habitability as the first step to achieve 'maximum' liveability for people' (Rueda, 2016, p.11). Urban habitability, the construct that implies optimizing conditions of urban life for habitants, enables the interaction between people and their environment.

Even though *Ecological Urbanism* is a brand-new concept for connecting people with nature and places, the human-oriented planning concept indeed exists for a long time, which can be traced back to Howard's *Garden City* concept. Since the middle of nineteenth century, the humanistic planning ideas gradually developed and were well known to the public; especially after the industrial revolution, with the accumulation of urban material resources. As a result, large number of people flowed into cities, and a series of social issues and conflicts appeared; for example, the overcrowding, disease and general misery in London where high population densities, poor housing and sanitary conditions caused miserable lives and early deaths. Due to people's uncomfortable feeling towards messy environment, the emergence of the Utopian concept became a reflection of people's dissatisfaction to the real world, which led some theorists to start exploring the humanistic planning ideas. From this point of view, the core value of *Garden City* theory can be seen as a conceptual and practical proposal for tackling negative effects of the unprecedented urbanization, especially during the industrial revolution and seeking for better city and better life—caring about people's basic needs and improving their living standards to achieve liveable human habitat.

*Environmental Psychology* is regarded as the study of transaction between people and physical settings. In this transaction, individuals can change the environment, and their experience and behaviours are influenced by the environment. Academic researches (Bell et al., 1984; Bonnes & Secchiaroli, 1995; Bechtel & Churchaman, 2002; Gifford, 2002; Lang, 1987) in terms of environmental psychology highlighted how this concept is included within the topic of behavioural science—how people perceive, feel, sense and interact with their surrounding environments. Research methods of the *Environmental Psychology* study adopt some content that is much related to disciplines of sociology and psychology, mainly focusing on the collection and analysis of spatial information. Although lacking specific ecological research from the perspective of urban planning, it analyses the influencing mechanism of people's behaviours and cognition in the space.

## References


Alexander, C. (1977). *A pattern language* (1st ed.). Oxford University Press.

Alexander, C., et al. (1987). *A new theory of Urban Design* (1st ed.). Oxford University Press.

Baker, R. A. (1968). The future of psychological instrumentation. *Behavioural Sciences, 13*(1), 1–17.

Barker, R. G. (1976). On the Nature of the Environment. Cited in: Proshansky, H. M., Ittelson, W. H., and Rivlin, R. G. (1976). *Environmental Psychology: People and Their Physical Settings*. 1st ed. New York: Holt, Rinehart and Winston. pp.12–26.

Beatley, T. (2000). *Green urbanism: Learning from European cities* (1st ed.). Island Press.

Bechtel, R., & Churchman, A. (2002). *Handbook of environmental psychology* (1st ed.). John Wiley & Sons.

Beck, U. (1992). *Risk society: Towards a new modernity* (1st ed.). Sage Publications.

Bell, A. P., Greene, T. C., Fisher, J. D., & Baum, A. (1984). *Environmental psychology* (1st ed.). Harcourt Brace College Publishers.

Bonnes, M., & Secchiaroli, G. (1995). *Environmental psychology: A psycho-social introduction* (1st ed.). Sage Publications, Inc.

Bronfenbrenner, U. (1979). *The ecology of human development: Experiments by nature and design* (1st ed.). Harvard University Press.

Caprotti, F. (2013). Critical research on eco-cities? A walk through the Sino-Singapore Tianjin Eco-City. *China. Cities, 36*(2014), 10–17.

Chokor, B. A. (1987). Research policy and review 13 environmental-behaviour-design research techniques: An appraisal and review of the literature with special reference to environmental and planning-related information needs in the third world. *Environment and Planning A, 19*(1), 7–32.

Churchman, A. (1993). A differentiated perspective on urban quality of life: Women, children and elderly. Cited In: Bonnes, M. (1993). Perception and evaluation of environmental quality. 1st ed. UNESCO Programme on Man and Biosphere. pp. 165–178.

Churchman, A. (2002). Environmental Psychology and Urban Planning: Where Can the Twain Meet? [Online]. Available at: https://www.researchgate.net/publication/284027102. [Accessed on 10th December 2022].

Corner, J., Mertins, D., & Najle, C. (2003). *Landscape urbanism: A manual for the Machinic landscape* (1st ed.). Architectural Association Press.

Cosco, N. G. (2006). Motivation to Move: Physical Activity Affordances in Preschool Play Areas [Dissertation]. *Edinburg College of Art (Scotland): School of Landscape Architecture*, Heriot Watt University.

Darley, G. (1975). *Villages of vision: A study of strange utopias* (1st ed.). Architectural Press.

Dobbins, M. (2009). *Urban Design and people* (1st ed.). John Wiley & Sons Inc.

Douglas, I., et al. (2012). *The Routledge handbook of urban ecology* (1st ed.).

Fainstein, S. S. (2010.) Social justice and ecological urbanism. Cited In: Mostafavi, M. & Doherty, G., (2010) Ecological Urbanism. 1st ed. E&B engelhardt und bauer. pp. 300–301.

Friedmann, J., & Hudson, B. (1974). Knowledge and action: A guide to planning theory. *Journal of the American Institute of Planners., 40*(1), 2–16.

Gattegno, N. (2011). Ecological Urbanism. *Architectural Theory Review., 16*(1), 73–75.

Gehl, J. (1987). *Life between buildings: Using public space* (2nd ed.). Van Nostrand Reinhold.

Gibson, E. J., & Pick, A. D. (2000). *An ecological approach to perceptual learning and development* (1st ed.). Oxford University.

Gibson, J. J. (1979). *The Ecological Approach to Visual Perception*. 1st ed. Houghton, Mifflin and Company.

Gibson, J. J. (1982). Notes on Affordances. Cited in: Reed, E. and Jones, R. (1982). Reasons for Realism: The Selected Essays of James J. Gibson. 1st ed. London: *Lawrence Erlbaum Associates Inc*. pp.401–418.

Gibson, J. J. (1986). *The ecological approach to visual perception* (1st ed.). Lawrence Erlbaum Associates Inc.

Gifford, R. (2002). *Environmental Psychology: Principles and Practices*. 2nd ed. Boston, MA: Pearson Custom Publishing.

Gifford, R. (2007). *Environmental psychology: Principles and practices* (4th ed.). Optimal Book.

Good, R. M., Rivero, J. J., Zitcer, A., Umemoto, K., Lake, R. W., Baum, H., Forester, J., & Harrison, P. (2017). Confronting the challenge of humanist planning/ towards a humanist planning/a humanist perspective on knowledge for planning: Implications for theory, research and practice. /to learn to plan, write stories/three practices of humanism and critical pragmatism/humanism or beyond. *Planning Theory & Practice, 18*(2), 291–319.

Griffith, M. (2013). *Ecological Urbanism.* [Online]. [Retrieved from 11th May 2021]. https://treehuggingurbanism.wordpress.com/wpcontent/uploads/2013/05/griffith_deepecologicalurbanism_spreads1.pdf

Hagan, S. (2015). *Ecological urbanism: The nature of the City* (1st ed.). Routledge.

Harding, S. (2015). *Objectivity and diversity: Another logic of scientific research* (1st ed.). University of Chicago Press.

Hardy, D. (1989). The Garden City campaign: An overview. Cited In: Ward, S. V. (1992.) the Garden City: Past, present and future. Spon, E & FN. pp. 187–209.

Heimsath, C. (1978). Behavioural architecture: Design as if people mattered. *Cornell Hospitality Quarterly., 19*(1), 40–48.

Hodson, M. & Marvin, S. (2010). Urbanism in the Anthropocene: Ecological Urbanism or Premium Ecological Enclaves? *Cities. 14*(3): 298–313.

Howard, E. (1902). *Garden cities of tomorrow* (2nd ed.). S. Sonnenschein Ltd.

Hugel, S. (2017). From Garden City to Smart City. *Urban Planning, 2*(3), 1–4.

Jacobs, J. (1961). *The death and life of great American cities* (1st ed.). Random House.

Jacobs, J. (1985). Downtown is for people. Cited In: Willian, H. & Whyte, J. R.. (1985). The exploding Metropolis. 1st ed. University of California Press.

Jones, A. (1996). The psychology of sustainability: What planners can learn from attitude research. *Journal of Planning Education and Research.*

Joss, S., Cowley, R., de Jong, M., Müller, B., Park, B.-S., Rees, W., Roseland, M., & Rydin, Y. (2015). *Tomorrow's City today: Prospects for standardising sustainable urban development* (1st ed.). University of Westminster.

Kidder, L., & Fine, M. (1987). Qualitative and quantitative methods: When stories converge. *New Directions for Program Evaluation, 35*, 57–76.

Lake, R. W. (2017). A humanist perspective on knowledge for planning: Implications for theory research and practice. *Planning Theory & Practice, 18*(2), 302–305.

Lang, J. T. (1987). Creating the Architectural Theory: *The Role of the Behavioural Science in Environmental Design.* 1st ed, New York: Van Nostrand Reinhold Company.

Lang, J., Burnette, C., Moleski, W., & Vachon, D. (1974). *Designing for human behaviour: Architecture and the behavioural sciences* (1st ed.). John Wiley & Sons, Inc.

Lye, L. F., & Chen, G. (2010). *Towards a liveable and sustainable urban environment* (1st ed.). World Scientific Publishing Ltd.

Lynch, K. (1960). *The image of the City* (1st ed.). MIT Press.

Lynch, K. (1972). *The openness of open space* (1st ed.). Arts of Environment.

Lynch, K. (1981). *A Theory of Good City Form.* 1st ed. Cambridge, Massachusetts: MIT Press.

Marinic, G. (2012). Book Review. *Design Issues, 28*(1), 105–106.
Matan, A., & Newman, P. (2016). *People City: The life and legacy of Jan Gehl* (1st ed.). Island Press.
McHarg, I. L. (1967). An ecological method for landscape architecture. *Landscape Architecture, 57*(1), 105–107.
Mostafavi, M., & Doherty, G. (2010). *Ecological Urbanism* (1st ed.). E&B engelhardt und bauer.
Mumford, L. (1968). *The City in history: Its origins, its transformations, and its prospects* (1st ed.). Mariner Books.
Neisser, U. (1967). *Cognitive psychology* (1st ed.). Appleton, Century & Crofts.
Parham, S. (2013). *Shaping Sustainable Urbanism: Are Garden Cities Answer?* [Online.] [Retrieved from 20th January 2018.] http://uh-sustainable.co.uk/docs/Shaping_sustainable_urbanism.pdf
Parsons, K. C., & Schuyler, D. (2002). *From the Garden City to green cities: The legacy of Ebenezer Howard* (1st ed.). Johns Hopkins University Press.
Pickerill, J. (2013). Building liveable cities: Urban low impact development as low carbon solution? Cited In: Bulkeley, H., Broto, V. C., Hodson, M. & Marvin, S.. (2013). Cities and low carbon transitions. 1st ed. Routledge. pp. 178–197.
Reed, E. S. (1996). *Encountering the world: Toward an ecological psychology* (1st ed.). Oxford University Press.
Rogers, R. (1998). *Cities for a small planet* (1st ed.). Westview Press.
Roseland, M. (1997). Dimensions of the Eco-City. *Cities.*
Rosenau, H. (2007). *The Ideal City: Its architectural evolution in Europe* (3rd ed.).
Rueda, S. (2016). *Ecological Urbanism.* Urban Ecological Agency of Barcelona. [Online]. [Retrieved from 2nd October 2018]. https://www.vitoria-gasteiz.org/http/wb021/contenidosEstaticos/adjuntos/es/80/86/38086.pdf
Sharifi, A. (2016). From Garden City to Eco-Urbanism: The Quest for Sustainable Neighbourhood. *Sustainable Cities and Society.* 20:1 16.
Spaargaren, G., & Mol, A. P. J. (1992). Sociology, environment and modernity: Ecological modernization as a theory of social change. *Society and Natural Resources, 5*(4), 323–344.
Spirn, A. W. (2000). Ian McHarg, landscape architecture and environmentalism: Ideas and methods in context. Cited In: M. Conan., ed. (2000). Environmentalism and landscape architecture. Dumbarton Oaks. pp. 97–114.
Spirn, A. W. (2014). Ecological urbanism: A framework for the Design of Resilient Cities. Cited In: Foster, O. N., The ecological design and planning reader. 1st ed. Island Press. pp. 557–596.
Town and Country Planning Association Website. (No date shown). *Garden City Principles.* [Online.] [Retrieved from 28th May]. https://www.tcpa.org.uk/garden-city-principles

Turnbull, J. (2010). *Ecological Urbanism Review.* [Online]. [Retrieved from 10th February 2023]. https://www.domusweb.it/en/reviews/2010/12/02/ecologicalurbanism.html

Waldheim, C. (2006). Landscape as Urbanism. Cited In: Waldheim C. (2006). *The Landscape Urbanism Reader.* 1st ed. New York: Princeton Architectural Press.

Waldheim, C. (2010). Weak Work: Andrea Branzi's 'Weak Metropolis' and the Projective Potential of an 'Ecological Urbanism' Cited in: Mostafavi, M. and Doherty, G. (2010). *Ecological Urbanism.* 1st ed. Germany: E&B engelhardt und bauer. pp.114–121.

Walters, P. (2005). *Knowledge in the making: Prototyping and human-centred design practice.* Sheffield Hallam University.

Ward, S. V. (2016). *The peaceful path: Building garden cities and new towns.* Hertfordshire Press.

Whyte, W. H. (1980). *The social life of small urban spaces* (1st ed.). The Conservation Foundation.

Withagen, R., & Michaels, C. F. (2005). On ecological conceptualizations of perceptual systems and action systems. *Theory and Psychology, 15*, 603–620.

Withagen, R., Poel, H. J., Araujo, D., & Pepping, G. J. (2012). Affordance can invite behaviour: Reconsidering the relationship between affordances and agency. *New Ideas in Psychology., 30*(2), 250–258.

Yu, K. J. (2010). Introduction of book ecological urbanism (Chinese vision). Cited In: Mostafavi, M. & Doherty, G. Ecological Urbanism. 1st ed. E&B engelhardt und bauer. pp. 2–3.

Yu, L. (2014). Low carbon Eco-City: New approach for Chinese urbanisation. *Habitat International, 44*(2014), 102–110.

CHAPTER 3

# Public Spaces, People's Needs, and Behaviours to Understand People-Place Relationship

## 3.1 Introduction

Relating to the people-oriented development path of Chinese *New-Type Urbanization*, this research is therefore a call to consider the humanistic dimension, especially the area of people's needs and behaviours, as the foundation on which future Chinese urbanization could be usually envisaged. As Caprotti (2018) revealed, the notion of people's needs requires 'a sufficiently granular definition that is both place-specific and can be related to broader agendas and contextualisation in place-specific geographical and other factors' (p. 3). In this research, *the humanistic planning under the context of ecology* refers to planning approach that prioritizes the natural environment and people's well-being. This approach regards cities not only as economic and technological systems, but also as ecosystems that interconnected with the natural world, which intends to play the role of advocacy trend of thought for Chinese eco-city's practices. Specific contribution of this research in relation to Chinese urbanism has been regarded as humanistic planning is conducive to promoting a more sustainable and people-oriented approach to urban development, as the rapid pace of Chinese urbanization has created lots of challenges which include environmental degradation and social inequality. *Humanistic planning under the context of ecology* seeks to address these



Y. Wu, *Possibilities and Limitations of Chinese Eco-City Development*, New Perspectives on Chinese Politics and Society,
https://doi.org/10.1007/978-981-95-1958-3_3

challenges by prioritizing people's needs and taking people-place interactive relationship into much more consideration during eco-city's planning and designing process. In other words, by considering long-term impacts of urbanization on the environment, humanistic planning helps to create more sustainable cities that are better equipped to address challenges about Chinese rapid urbanization process, especially during the period of urbanization transition.

Designers are claimed to be designed for people and to be interested in designing an environment that can 'uplift the spirit and enhance the sense of well-being' (Lang et al., 1974, p. 3). People usually perform their activities based upon how they perceive their built environment, therefore, it requires planners and designers to create a living environment that can communicate with people. However, people's needs and their behaviours are diverse and are changing from one person to another, so in order to meet people's needs, it is of great importance to be informed of this changing and diversity (Churchman, 2002; Lang, 1987) during eco-city's planning process.

## 3.2 Rational for Choosing Public Spaces, People's Needs and Behaviours to Understand People-Place Relationship

*Garden City* as a way to provide the classic theoretical foundation for the people-place relationship that supports eco-city's planning and design towards the humanistic direction. To make this conceptual relationship more concrete and much suitably applied into eco-city project, this research chooses public space as the representation of *place*, people's needs and their behaviours as the representation of *people*.

On the one hand, in this research, the *place* refers to *public spaces* within the eco-city, as they typically provide the physical environment for a variety of people's outdoor activities, including 'walking, biking, driving, riding, parking, getting on and off transit, going in and out of buildings, sitting, dining, picnicking, hanging out, getting together, playing, relaxing, having festivals, partying' (Dobbins, 2009, p.11). People's activities often occur in these places, such as sidewalks, streets, parks, plazas and public squares. A typology of urban open spaces has been shown in Table 3.1. Public spaces and people's activities within these areas have represented the connective relationship—public spaces contribute to connecting people with each other and also with their activities. As Dobbins stated (2009), public spaces and people's activities that 'frame them make up the physical component of what gives places their character, their memorability and their identity' (p.11). On the other hand, no matter in industrial design or in computer science discipline, the core ideology of people-oriented principle refers to putting people at the heart of design process and meeting users' needs. In addition, people's needs would seem to be the 'underlying causes of people's behaviours and behavioural changes. Without any need there would be no motivation to perform a behaviour' (Vlek, 2000, p.160). Thus, people's behaviours are regarded as the manifestations of people's higher-level needs in this research.

## 3.3 Value of Public Spaces in Sustainable Social Development

Public spaces, in different parts of human life, play a vital role (Kurniaty, 2014, p. 507). Allocation and efficient use of them is indeed a fundamental issue that should be taken into consideration in studying people's lives (Garcia-Ramon et al., 2004, p. 215). The use of urban public spaces is an experience that is not similar for each person to understand urban life due to a variety of reasons, such as age, gender, ethnic-racial minorities (Rafieian & Khodaei, 2009, p.230). In order to achieve factors' role in impacts of city on people, understanding 'the impact of urban public space on the networks of social relations is required' (Fathi, 2012, p.50), which has been seen as the foundation of people-place relationship put forward in this research. Thus, design features of urban public spaces are significant factors not only promoting social activities, but also creating an environment which 'comes along relations, meaning and purpose, also constructs the basis for sustainable social development' (Abada et al., 2007, p. 2011).

Public places have existed since people began making their settlements. To some extent, public spaces can be regarded as the house of all as

**Table 3.1** A Typology of Urban Open Spaces

| *Type of Open Space* | *Characteristics* |
|---|---|
| Public Parks | |
| Public/Central Park | Publicly developed and managed open space as part of zoned open space system of city; open space of city-wide importance; often larger than neighbourhood park. |
| Neighbourhood Park | Open space developed in residential environments; publicly developed and managed as part of the zoned open space of cities, or as private residential development. |
| Vest-pocket Park (Community Park) | Small urban park bounded by buildings; may include fountain or water feature. |
| Market | |
| Open space or streets used for markets. | |
| Streets | |
| Pedestrian Sidewalks | Part of cities where people move on foot; most commonly sidewalks and paths, planned or found, which connect one destination with another. |
| Playgrounds | |
| Play area located in neighbourhood, includes traditional play equipment such as slides, amenities for adults such as benches. | |
| Community Open Spaces | |
| Community Garden/Park | Neighbourhood spaces designed or managed by local residents on vacant land; may include viewing gardens, playing areas and community gardens, not officially viewed as part of open space system of cities. |
| Greenways | Interconnected recreational and natural areas connected by pedestrian and bicycle paths. |
| Indoor Marketplaces | |
| Marketplace/Shopping Centre | Interior and private shopping areas, may sometimes be called 'festival marketplaces'; privately developed and managed as part of office or commercial development. |
| Neighbourhood Spaces | |
| Everyday Spaces | Publicly accessible open places such as street corners and steps to buildings, which people can use. |
| Neighbourhood Spaces | Publicly accessible open space, such as street corners and lots near where people live; can also be vacant or undeveloped space located in neighbourhood including vacant lots and future building sites; used by children, teenagers and local residents. |

(*Source* Information is from Francis. 2003. pp. 6–7.

they give people a ticket to the physical environment, and people create these spaces to support their activities that occur in the public realm. In a city, people usually use the set of social functions of public space for their daily activities, such as sports, games, chatting and economic exchange. However, where urban space is used mainly for motorized mobility (which is one of functions for public space among others, but not the main one), the resident becomes a mere pedestrian. From this perspective, public spaces need to have their own potential uses and functions, usually flexibly across time and space, and forms that public places tend to work best or to be human-centred when they 'reflect the nature of activities they support, like walking, gathering, sitting, transit, and parking' (Dobbins, 2009, p. 42).

In the post-pandemic era, public spaces, with the aim of moving towards the healthy and comfortable urban environment, gradually becomes the focus point of sociologists, geographers and urban planners. They are considered as social spaces which are 'open and accessible to anyone, regardless of people's gender and age' (Rafieian et al., 2021, p.34). Most studies have been done to assess the condition of public spaces (Sanei et al., 2017, p.243), however, the following experimental studies have been focused on the role of public spaces in sustainable social development. Turel et al. (2007) pointed out vital problems of urban public spaces are lack of walking possibility, insecurity, repairs and maintenance failure (pp. 2038–2041). Perovic and Kurtovic (2012) examined social functions of urban public spaces as counted security, availability, regularly organizing, calmness. And characterized the lack of attractiveness, lack of visual beauty and visual disturbance, as the worst indicators (p.926–928). Also, Fathi (2012) arrived to the conclusion that sustainable social relations are the product of sustainable urban spaces (p. 58). As Table 3.2 shows, different criteria of public spaces have been defined by some well-known theorists of this field.

Due to its multi-functional and multi-disciplinary nature, public space offers a holistic view on the city, including social inclusion, urban governance, urban economy and more intensive adaptation of the city to people's green and low-carbon lifestyle. To a certain extent, public space is essentially regarded as urban advantage, influencing the overall impression of a city as well as helping to increase its general prosperity. Thus,

**Table 3.2** Indicators of Urban Public Spaces

| *Theorist* | *Literature* | *Criteria* |
|---|---|---|
| Lynch, K. | The openness of open space. 1972. | Compatibility, access and control, efficiency, equity, social vitality. |
| Bentley, I. | Responsive environment: A manual for designers. 1985. | Permeability, variety, flexibility, permeability, variability (form), readability, flexibility, visual adaptability, richness, possibility of personalizing. |
| Trancik, R. | Finding lost space: Theory of Urban Design. 1986. | Keeping the movement connections, focusing on spaces, continuity of the edges, controlling the axis of vision and perspective, combining of the internal and external spaces. |
| Punter, J. V. and Carmona M. | The design dimension of planning: Theory, content and best practice for design policies. 1997. | Land use, pedestrian movement, patterns of behaviour, riding movement flow, permeability, public perception, qualitative evaluations, operations, concept recalling, objective perspective, synthetic form. |
| Pakzad, J. | Development of thoughts in urbanization: From quantity to quality. 2007. | Safety, vitality, space, flexibility. |
| Habibi, S. M. | From town to City: A historical analysis of the meaning of the City and its form. 2010. | Diversity, safety and security, compatibility, social role, distinction and character, legibility, permanence, spatial contrast, turning points, pausing places, human scale, flexibility, weather conditions. |

(*Source* Information is summarized by the researcher.).

it is essential for a city to recognize the importance of public space and the need for new approaches in its creation and maintenance. As Gažová (2011) illustrated the impact and significance of public spaces on the landscape—'[public space] reflects in all areas of urban life, and it is necessary to apply an environmental approach as a basis and starting point for processes in urban spatial planning' (p.10).

According to Scopelliti et al. (2015), open spaces in cities afford a variety of opportunities for their inhabitants and visitors. Regeneration or planning of public spaces can offer psychological, social (Tilt et al., 2007) and aesthetic benefits. Also, public spaces in any forms create an opportunity for social interaction and thus represent the largest investment to increase social capital (Tilt et al., 2007). Quality of public spaces can have significant impacts on the economic condition of cities, and should therefore be an integral part of any successful regeneration strategy. For instance, Barton and Pineo (2009) demonstrated the pleasant, lively and well-maintained public space contributes to increase the number of visitors. Also, Lawlor et al. (2018) illustrated case studies have shown quality public spaces increase people's interest and attract inhabitants to the site.

## 3.4 People's Needs and Behaviours Matter

The concept of human needs has been often mentioned as a mental or cultural notion by different subjects, such as economics related to consumer demand, sociology related to social psychological needs. As Zeisel (1974) illustrated, needs of users should be defined regarding the 'underlying social meaning of behaviour and perception to the user group, rather than merely in terms of what users say they need' (p.293). The well-known model for illustrating human needs is the one proposed by Abraham Maslow (1943) in his paper *A Theory of Human Motivation*. He used the five-level pyramid model (*See* Fig. 3.1) to reveal kind of things that human needs included—physiological needs are on the bottom of this hierarchy followed by safety needs, social needs and esteem needs, which are all deficiency needs (*D-Needs*). The top level is self-actualization, which has been classified as growth needs (*G-Needs*). 'Maslow later divides the growth needs of self-actualization, specially adds two lower-level growth needs, namely cognitive needs and aesthetic needs, prior to general level of self-actualization and one beyond that level, that is self-transcendence, which is sometimes referred to human spiritual needs' (Zhang & Dong, 2008, p.4).

According to this model, Maslow believes people's behaviours in certain moments have been affected by the most severe needs. That is, when need satisfaction begins, a change will occur in people's motivations, then another high level of need will be more attractive than the previous one and will drive people's behaviours to pursue more. Needs will escalate until they reach the peak of the need hierarchy, then diminish once satisfied, shifting focus to other needs.

Study about people's behaviours contributes to offering a new research perspective to understand people's 'hidden' needs and can address eco-city's social and environmental issues from the angle of residents at the micro-scale. People's behaviours are seen as the manifestations of higher-level needs. To step into a new stage of people-centred design,

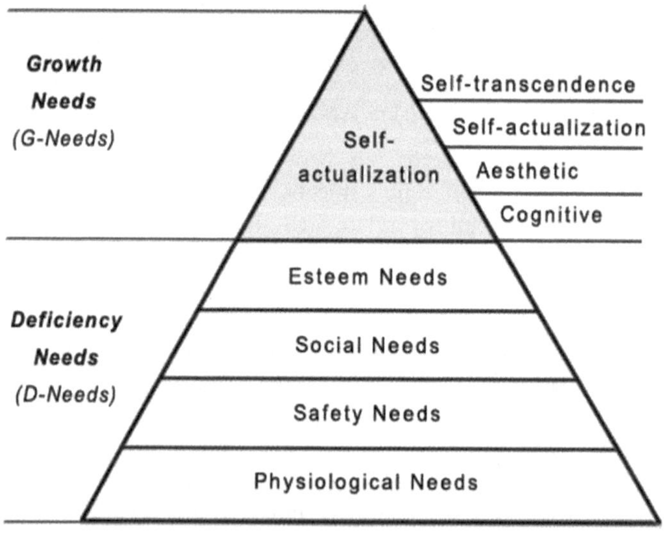


**Fig. 3.1** Maslow's Hierarchy of Needs Model (*Source* Figure is from Zhang & Dong, 2008, p.3)

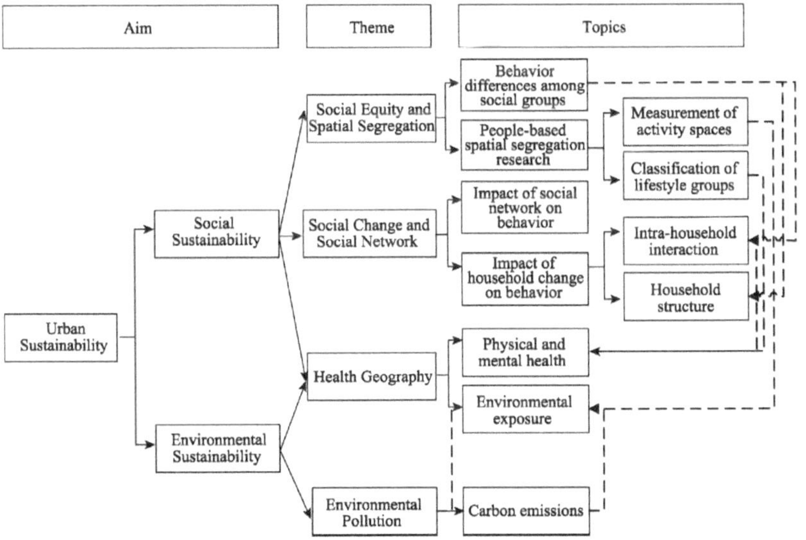


**Fig. 3.2** New trends in Chinese behavioural geography (*Source* Table is from Chai et al., 2014, p.1254)

this research argues the eco-city should find 'usable' solutions, more importantly, in order to put the people-oriented principle into planning practices, not just like a slogan, but eco-city planning ought to satisfy higher levels of people's needs. However, as shown in Table 3.4, this information usually cannot be easily caught by people's own words, their 'visible' daily behaviours to a large extent can reveal people's 'invisible needs'.

Urban planning can establish guidelines that may impact people's behaviours. For instance, in most Chinese eco-cities, the low-carbon

**Table 3.3** Examples of higher levels of people's 'invisible' needs

| *Maslow's Model* | *Examples* | *Needs Related to Eco-City* |
|---|---|---|
| Social needs | Belonging feeling | Community engagement |
| Esteem needs | Feeling of respect | Living needs of lower incomers |
| Self-actualization | Consciousness | People's sustainable lifestyles |

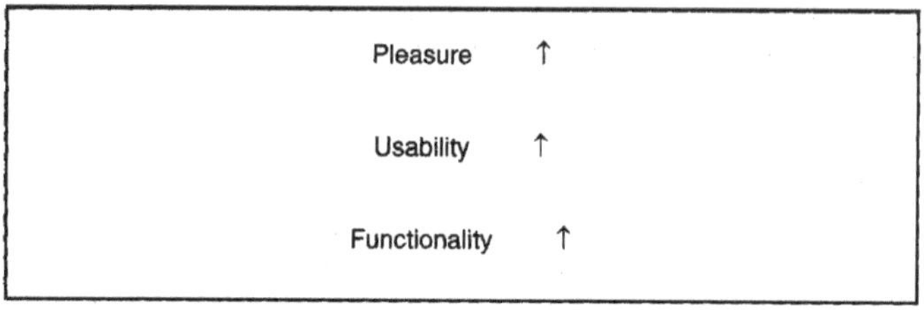


**Fig. 3.3** Jordan's Hierarchy of Consumer Needs Model (*Source* Figure is from Jordan, 2000, p.6)

target has been quantified by specific index system to promote people's pro-environmental behaviours. People follow them intuitively and planners can render them in designing processes. Currently, behavioural research in China has been mainly within geography discipline, using quantitative method to explore people's activities and their spatial pattern. As Chai et al. (2014) criticized, while Chinese urban geographers have made more methodological advances in data collection and analysis tools, 'theoretical development lags behind compared to the space-time behaviour approach in the West' (p.21). In terms of urban sustainability issues from behavioural perspective, four major new researches—social equity and spatial segregation, social change and social network, health geography and environmental pollution, have been paid much attention than ever before by Chinese geographers (*See* Fig. 3.3). Unlike the above-mentioned behavioural researches, this research makes the case for 'humanizing' newly built eco-city by emphasizing on social sustainability, through the lens of people's living experiences, particularly their outdoor activities in public spaces. It plans to argue that behavioural research, the core element of people-place relationship, offers a new perspective for understanding the complexity and diversity of people's hidden higher-level needs.

## 3.5 New-Type Conceptual Model to Measure Human-Centred Design

*Human-Centred Design (HCD)* is a design and management framework that develops solutions to problems by involving people's perspectives in all steps of problem-solving. This concept starts in ergonomics and human-computer interaction fields. Moreover, in the field of industrial product design, this concept has been identified in various terms (*See* Table 3.5), such as person-centred design or user-centred design, but similar characteristics can be summarized as putting people's (users') needs at the heart of design process and making things useful, usable and desirable for people. As Black (2006) pointed out, 'the central premise of user-centred design is about the best-designed products and services result from understanding the needs of people who will use them'.

Zhang and Dong (2008) used the *Design and Society Model* as a framework to illustrate the design history since the second half of twentieth century. They critically explored how a design found its own characteristics in five periods based upon the level of welfare and social view (*See* Fig. 3.4).

According to this modified *Design and Society Model*, human-centred design has always been the focus of attention for designers even though it has different meanings in different contexts.

- **Prior to the 1950s:** design put more emphasis on function. *Form Follows Function*, the typical slogan, reflected social phenomenon that if functioned well, a product is considered as meeting people's needs.

**Table 3.4** Different definitions of human-centred design

| *Authors* | *Terms* | *Definition* |
|---|---|---|
| Jordan (2000) | Person-centred design | To take a wider view of person-centred design in a holistic content. |
| Alison Black (2006) | User centred design | To engage with users to gather insights that can drive design from the earliest stages of product and service development. |
| Walters (2005) | Human-centred design | An exploration of human needs and aims to improve quality of life. |

- Putting people at the heart of design process.
- Making things useful, usable and desirable for people.

(*Source* Information is summarized by the researcher.)

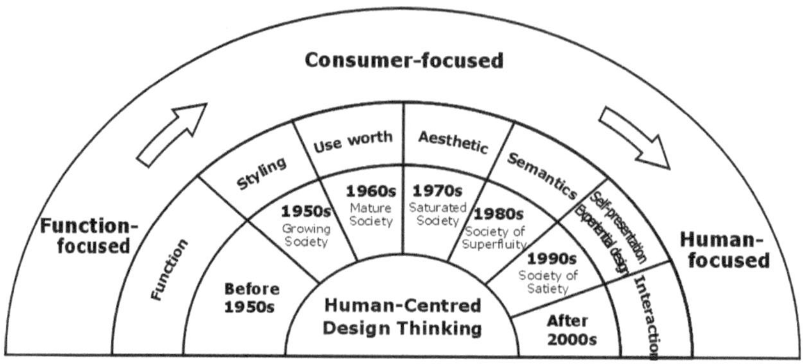


**Fig. 3.4** Modified 'Design and Society' Model (*Source* Figure is from Zhang & Dong, 2008, p.4)

- **1950s–1980s:** due to the industrialization and commercialization, design tended to be consumer-focused. Main tendencies during this period included 'appearance design, usage value, aesthetics and semantics. Although some phenomena like planned obsolescence focused on consumption, consumers' needs and aspiration drove human-centred thinking in this period' (Zhang & Dong, 2008, p.4).
- **1990s–present:** design paid more attention to human beings rather than consumers or users as the concept of human-centred design is gradually growing day-by-day. It is worth pointing out that the human being is different from a consumer or a user here, because people are much conceptualized as users or consumers. The distinction between these terms is important because it highlights the multifaceted nature of people's identity and roles. While a person can be a consumer and a user in various contexts, these roles do not define the entirety of their being. Human beings have broader aspirations, emotions, and connections that go beyond their roles as consumers or users. 'Researches on abilities and human-oriented ergonomic designs have emerged. Human-computer interaction has begun to influence other design field' (Zhang & Dong, 2008, p.4).

By combining the *Maslow's Hierarchy of Needs Model* (from the micro and psychological angle) and the *Design and Society Model* (from the macro and social angle), Zhang and Dong (2008) created a brand-new conceptual model which has bridged human needs and product design evolution (*See* Fig. 3.5). This model reveals the product design evolution, from functionality and usability to the higher-level desirability, which has responded to the hierarchy of human needs. And it argues that to enter the human-focused design phase, future design ought to satisfy different range of human needs, even some subtle needs which users have not recognized or unexpressed. Specific interpretation of this model is shown in Table 3.6.

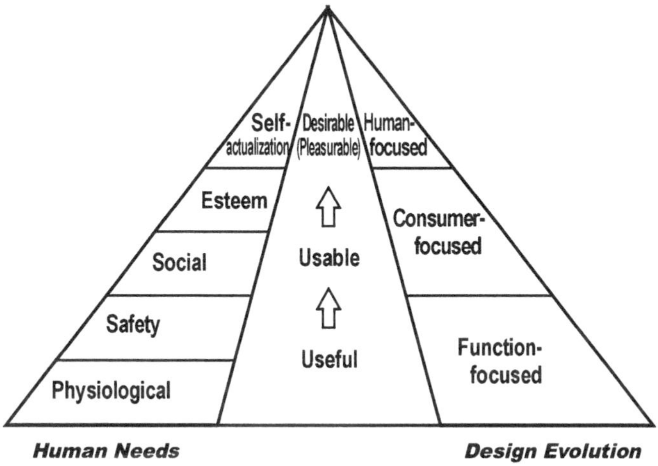


**Fig. 3.5** New-Type Conceptual Model of Human-Centred Design (*Source* Information is from Zhang and Dong, 2008, p.5)

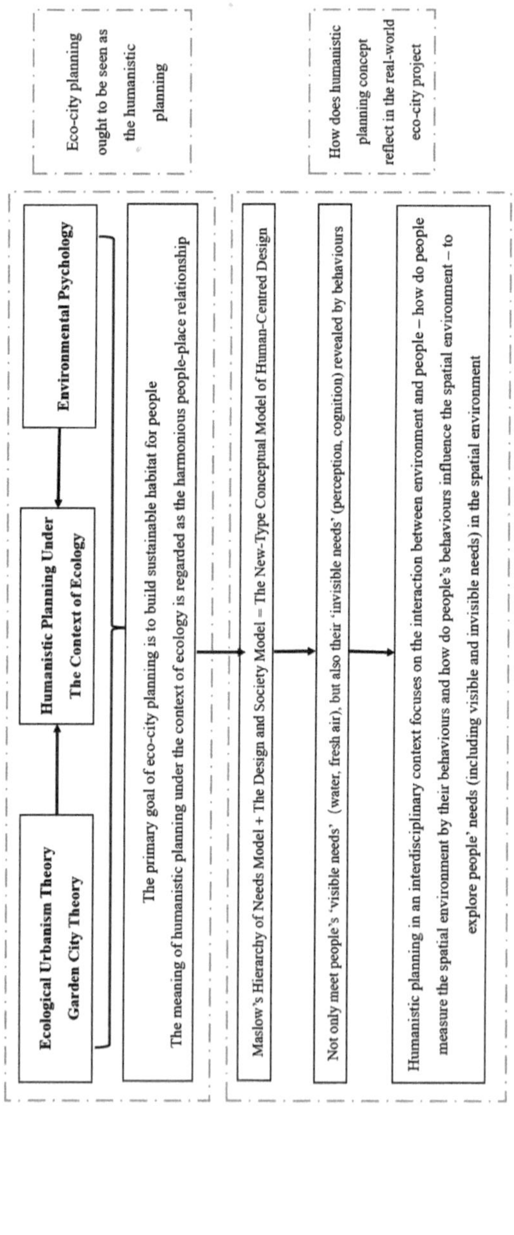


**Fig. 3.6** From literature review to analytical framework (*Source* Information is summarized by the research)

**Table 3.5** Interpretation of the Model of Human-Centred Design

| *Different Period* | *Meaning of Design* |
|---|---|
| Function-focused period | Design means producing useful products, which mainly meets the fundamental level of human needs, such as physiological and safety needs. |
| Consumer-focused period | Design develops to find usable solutions. Both social and esteem needs are taken into consideration during the designing process. |
| Human-focused period | Design intends to satisfy higher level of human needs, not only about the well thought-out of product's function and usability, but also about the cognitive or aesthetic aspects, in order to reveal the future's design tendency. |

(*Source* Information is from Zhang & Dong, 2008, p.5.)

## 3.6 Conclusion

The concept of human needs has been deemed as a relatively mental and cultural notion by different subjects, such as economics, sociology and philosophy. The well-known model for illustrating human needs is the one proposed by Abraham Maslow in his paper *A Theory of Human Motivation* (1943). He used the five-level pyramid to reveal people's needs not only include physiological needs but also include psychological ones. And Maslow believes people's behaviours in certain moments have been affected by the most severe needs. *The New-Type Conceptual Model of Human-Centred Design* reveals the product design evolution, from functionality and usability to desirability, which has responded to the hierarchy of human needs. Also, this model argues that to enter the human-focused design phase, design ought to satisfy different range of human needs, even some subtle needs which users have not recognized or unexpressed.

This research argues for a foregrounding of human needs at the heart of eco-city's societal futures. Although economic imperatives and technological generations are the focus of Chinese policy-making under the traditional urbanization development path, it is now time to place basic people's needs at the centre of thinking and planning for current eco-city's development under the green urbanization path. Under the traditional development model, urban future of eco-city development is not primarily a technical problem, nor is it a purely economic one, although it has elements of both. How to explore eco-city becomes the new highland for green innovation and the new model of liveable place between people and nature will be the priority of Chinese eco-city development.

In summary, this chapter focuses on the transition from the literature review to analytical framework. By introducing the new-type conceptual model, this chapter argues that people's needs matter because it is directly related to the people-centred design principle, and it also argues people's behaviour is the manifestation of higher-level (invisible) needs. By combining Chapter 2 and Chapter 3, it constructs the analytical framework based upon the existing literature review to determine the definition of people-oriented concept of this research—humanistic planning in an interdisciplinary context focuses on the interaction between people the environment (e.g. people measure the spatial environment by their behaviours), and its main objective is to explore people's needs (including visible and invisible ones) when they stay in the spatial environment. It is worth mentioning that so far before the next chapter, *Objective One* has been addressed—to review theories relating to the *humanistic planning under the idea of ecology* from three different disciplines and to explore connections among these discourses, for building an analytical framework responding to the research gap, while also offering theoretical support for further research analysis and also to review the new-type conceptual model related to human-centred design.

## References


Abada, T., Hou, F., & Ram, B. (2007). Racially mixed neighbourhood, perceived neighbourhood social cohesion, and adolescent health in Canada. *Social Science and Medicine, 65*, 2004–2017.

Barton, S. & Pineo, R. (2009). Human benefits of green spaces. Cited In: Barton, S. & Pineo, R. (2009). Sustainable landscapes series. 1st ed. University of Delaware.

Black, A. (2006). UK Design Council on User-Centred Design and Experience Design. [Online]. Available at: https://blog.experientia.com/ukdesign-council-on-user-centred-design-and-experience-design/ [Accessed on 19th April 2018.]

Caprotti, F. (2018). Future cities: Moving from technical to human needs. *Palgrave Communications, 4*(35), 1–4.

Chai, Y. W., Liu, Z. L., & Zhang, Y. (2014) Space-time behaviour research and application in China. Cited In: Kwan, P., Richardson, D., Wang, D. G. & Zhou, C. H. (2014). Space-time integration in geography and GIScience: Research Frontiers in the US and China. 1st ed. Springer. pp. 21–38.

Churchman, A. (2002). *Environmental Psychology and Urban Planning: Where Can the Twain Meet?* [Online]. Retrieved from 10th December 2022. https://www.researchgate.net/publication/284027102

Dobbins, M. (2009). *Urban Design and people* (1st ed.). John Wiley & Sons Inc.

Fathi, S. (2012). The analysis of social relations in sustainable urban space. *Journal of Iranian Social Development Studies, 4*, 47–64.

Francis, M. *Urban open space: Designing for user needs* (1st ed.). Island Press.

Garcia-Ramon, M. D., Ortiz, A., & Prats, M. (2004). Urban planning, gender and the use of public space in a peripheral neighbourhood of Barcelona. *Cities, 21*, 215–223.

Gažová, D. (2011). Urban-landscape contexts of creation and planning of urbanized landscape (in Slovak). *Urbanita, 23*(4), 8–13.

Habibi, S. M. (2010). *From town to City: A historical analysis of the meaning of the City and its form*. Tehran University Press.

Jordan, P. (2000). *Designing pleasurable products: An introduction to the new human factors* (1st ed.). Taylor and Francis.

Kurniaty, R. (2014). Local elites and public space sustainability: The local elite roles in the presence and usage of public space in Malang Raya, Indonesia. *Environmental Sciences, 20*, 506–515.

Lang, J. (1987). *Creating the Architectural Theory*. 1st ed. New York: Van Nostrand Reinhold Company.

Lang, J., Burnette, C., Moleski, W., & Vachon, D. (1974). *Designing for human behaviour: Architecture and the behavioural sciences* (1st ed.). John Wiley & Sons, Inc.

Lawlor, E., Tasker, M., Lee, R. & Hay, S. (2018). The pedestrian pound: The business case for better streets and places.

Lynch, K. (1972). *The openness of open space* (1st ed.). Arts of Environment.

Maslow, A. H. (1943). A theory of human motivation. *Psychological Review, 50*(4), 370–396.

Pakzad, J. (2007). *Development of thoughts in urbanization: From quantity to quality* (1st ed.). New Towns Civil Company.

Perovic, S., & Kurtovic, F. N. (2012). Visual perception of public open spaces in Niksic. *Social and Behavioural Sciences, 68*, 921–933.

Punter, J. V., & Carmona, M. (1997). *The design dimension of planning: Theory, content and best practice for design policies* (1st ed.). Spon Press.

Rafieian, M., Azimi, M. & Moghaddam, H. (2021). Quality Evaluation of Urban Zones in Success of Public Spaces by the Use of Veroni Weighted Diagram (VWD): A Case Study of Lar Old City. *Amayesh Journal*. 5:33–49.

Rafieian, M., & Khodaei, Z. (2009). Evaluation criteria and standard affecting the satisfaction of citizens in urban public spaces. *Journal of Rahbord University, 53*, 227–248.

Sanei, M., Khodadad, M. & Ghadim, F. P. (2017). Effective instructions in design process of urban public spaces to promote sustainable development.

Scopelliti, M., et al. (2015). Go greener, feel better? The positive effects of biodiversity on the well-being of individuals visiting urban and Peri-urban green areas. *Landscape and Urban Planning., 134*, 221–228.

Tilt, J. H., Unfried, T. M., & Roca, B. (2007). Using objective and subjective measures of neighbourhood greenness and accessible destinations for understanding walking trips and BMI in Seattle, Washington. *American Journal of Health Promotion, 21*(4), 371–379.

Trancik, R. (1986). Finding lost space: Theory of Urban Design. *Van Nostrand Reinhold, 74*, 32–38.

Turel, H. S., Yigit, E. M., & Altug, I. (2007). Evaluation of elderly people's requirements in public open spaces: A case study in Bornova District (Izmir, Turkey). *Building and Environment, 42*, 2035–2045.

Vlek, C. (2000). Essential psychology for environmental policy making. *International Journal of Psychology, 35*(2), 153–167.

Walters, P. (2005). *Knowledge in the Making: Prototyping and Human-Centred Design Practice*. Sheffield Hallam University.

Zeisel, J. (1974). *Sociology and user analysis research* (1st ed.). Russell Sage Foundation.

Zhang, T. & Dong, H. (2008). Human-Centred Design: An Emergent Conceptual Model. In *Royal College of Art*, April 8–10 2009.

CHAPTER 4

# Within-Case Analysis of the Humanistic Planning Concept in Chinese Eco-City Development

## 4.1 Introduction

From the perspective of urban planning, indicator system of the SSTEC has continuously shown the ongoing trend of people-oriented development pattern in these years since planning and construction stages. From the perspective of urban design, the masterplan has been formulated to provide a clear blueprint to guide the SSTEC's overall urban development to achieve its KPIs, and to offer people a life-convenient living settlement under the general people-oriented design concept. In the scope of the second half of this chapter, it plans to introduce elements that could constitute responsive public spaces of the SSTEC, including permeability and legibility as well as variety, to argue that people-oriented concept has been rooted in the level of concrete urban design means.

## 4.2 Emerging Eco-City Concept in China and Planning System to Guide Eco-City Development

It is obvious China has devoted significant political will and economic resources to the development of newly built eco-city planning projects, reflecting government's goal to build the harmonious society in which



Y. Wu, *Possibilities and Limitations of Chinese Eco-City Development*, New Perspectives on Chinese Politics and Society,
https://doi.org/10.1007/978-981-95-1958-3_4

environmental sustainability and social stability are mutually reinforcing. Under such development trend, existing issues and potential obstacles of Chinese ecological urban planning as well as its development possibilities and reform directions are worthy of attention for promoting eco-city's construction. This section aims to give an overview of emerging eco-city concept and related policies to contextualize planning of the SSTEC.

The meaning of planning incorporates a variety of activities and is defined as 'an activity of some considerable breadth, both in theory and practice' (Adam & Watkins, 2014, p. 14). Such description contributes to understanding urban planning and to thinking about its broader purposes, rather than paying much attention to planning methods. In the Chinese context, urban planning is widely seen as 'a tool that can facilitate growth' (Wu et al., 2015, p. 2). The rationale of growth-oriented planning in China provides evidence that 'planning by its nature and a wide spectrum of associated activities can be supporting growth' *(ibid.)*. Reviewing appropriately 50-years' urbanization development since the reform and opening-up, China constantly explores the path of urbanization suitable for its development pattern by reform, which has been seen as the key to obtain growth achievements (Su & Wei, 2018, p. 50). However, its rapid rate and sheer scale has been increasingly pressing problems—due to some typical urbanization issues, such as large population with relatively little land, a huge population base and obvious urban-rural gap, China has embarked on the development path of *New-Type Urbanization* (Wei, 2014).

Given urban issues China has to face in present and future, it is vital for China to transform towards a more sustainable, people-oriented and high-quality form of urbanization. The *3rd Plenary Session of the 18th CPC Central Committee* proposed the major policy of *sticking to the path of New-Type Urbanization with Chinese characteristics, and pushing forward people-oriented urbanization* (坚持走中国特色新型城镇化道路，推进以人为核心的城镇化). Following this planning principle, *The National New-Type Urbanization Plan (2014–2020)* was released by the central government in early 2014, and formally issued to mark the major transformation of Chinese urbanization (Lu & Chen, 2015). Under such transformative trend, the proliferation of eco-city development comes at a time when China is reaching peak urbanization rate, making most of cities being a global trend in which considerations of ecological and low-carbon development have been widely taken into both urban governance and city construction. As Yu (2014) pointed out, concepts of *eco-city*, *low-carbon city* and *low-carbon eco-city* are adopted to cope with risks and pressure of these social and environmental issues, which has been regarded as 'a new approach to urbanisation in China' (p. 102). And planning framework to support Chinese eco-city development is seen as two central government's mechanisms for promoting the ecological human settlement development.

> [One] is to award demonstration projects and best practices to encourage other cities to follow their experiences and principles. ... [Another] is to compose relevant policies by circulating government documents, such as defining carbon emission targets for each provincial and municipal government, or encouraging low carbon, energy efficient or environmental protection projects and programmes with appropriate financial support to the awarded programmes or projects. This mechanism requires coordination among various central ministries. (Yu, 2014, p. 105)

Although the idea of adopting ecological principles in urban planning surfaced in China in the late 1980s, it is only in the last decade that eco-city, functioned as an urban development model, was considered and then put into action. Since the early 1990s, Chinese central institutions, particularly the *Ministry of Environmental Protection (MEP)* and the *Ministry of Housing and Urban-Rural Development (MoHURD)*, attempted to guide cities towards a more sustainable development path by formulating low-carbon policies and performance indicators. As early as in the year of 1995, the *Ministry of Environmental Protection (MEP)* issued guidelines for building the demonstration eco-communities and environmental protection model cities under the eco-construction programme (Liu, 2009, p. 251).

Since it was formally adopted by the central government at the *17th National Congress of the Communist Party of China* in 2007, China has chosen to frame the eco-city concept within the utopian ideal of an *eco-culture* which calls for building the harmonious society. Moreover, many local governments have begun to develop eco-cities on their own or together with international partners. According to surveys undertaken by *Chinese Society of Urban Studies*, 'around 97% of prefectural cities ... had announced or started eco-city or low-carbon city as their development strategy by 2012. A sample survey among these cities shows that 53% of them had already been under their process of different scales and types of ecological urban development, 28% were inviting planning consultants to produce eco-city plans, while 19% were devising their strategic policies for making for eco-city development' (Yu, 2014, p. 103). Such survey has been released to obtain feedback from different parties on their perspective on KPIs for an eco-city in China. 'The eco-city index system to be developed is intended as a measurement tool to guide and measure Chinese eco-city development' (Asian Development Bank, 2010, p. 1).

The current Chinese planning system was legalized in the *City and Rural Planning Act 2008*. As the improved version of the 1989 Act, the new feature of this planning system is about the introduction of master plans and detailed plans for towns and villages. In addition, as Wu (2007) pointed out, the *City and Rural Planning Act 2008* proposed strategic development plans or outline strategies (concept plans) to bridge the gap between system plans and specific master plans. However, the *2008 Act* did not legalize urban design as part of the planning system. 'Alongside the updated system of plans, planning policies were also developed into a complex structure, regulating not only plan-making, but also the administrative responsibilities of the parties involved in urban development, plan implementation and monitoring' (Chen, 2016a, 2016b, p. 83). As Fig. 4.1 shows, planning policies include ordinances (条例) published by the state council, local by-laws (条例) published by provincial and municipal governments, administrative regulations (管理办法) published by the ministries or local authorities, ministerial and local advices or notices (通知). Local planning policies aim to respond to specific development conditions, and to enforce the decentralized decision-making mechanism. The SSTEC project focuses on design principles covered in the operation of this planning system.

## 4.3 Sino-Singapore Tianjin Eco-City: The Flagship Planning Project Between China and Singapore

It is of great importance to delineate some general information about the SSTEC project at the beginning of this section—all basic but unique characteristics can affect further data collection and analysis of this research.

The SSTEC has been proposed to against the backdrop of Chinese rapid urbanization and to promote eco-urbanism as a new model for sustainable development, in line with the emerging political agenda of *Constructing Ecological Civilization* (生态文明建设). This national-level

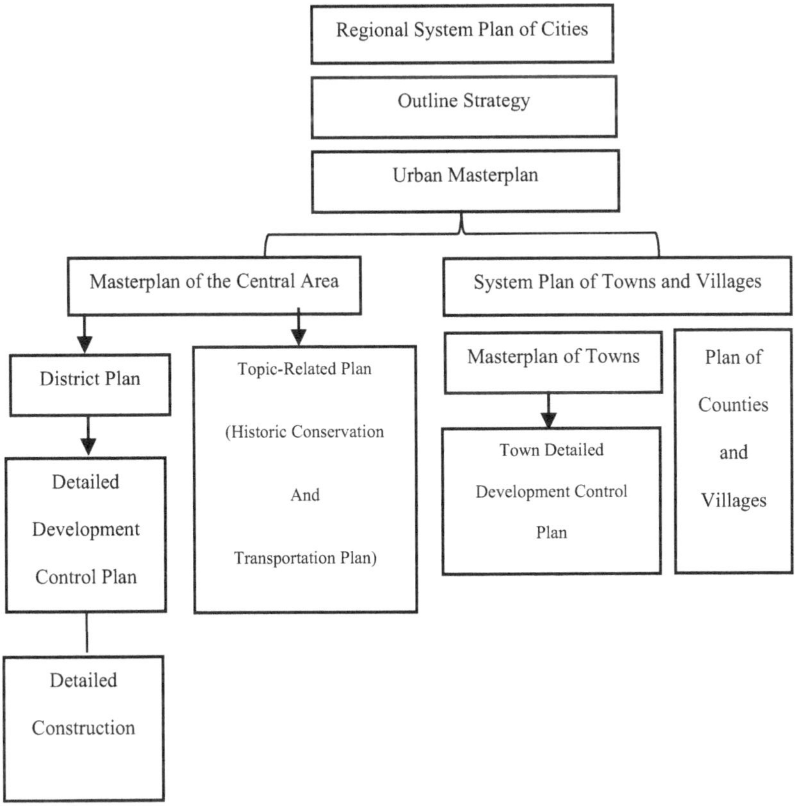


**Fig. 4.1** Legal plans in the planning system. *Source* Image is from Chen (2016a, 2016b, p. 83)

planning project plans to integrate Singaporean-style high-rise residential blocks with green belts and areas for the diverse economic activities. Three special project features have made the SSTEC be a representative case of Chinese eco-city project. Firstly, being recognized as the government-to-government flagship project collaboration between China and Singapore, it was reclaimed from the wasteland into economically productive and habitable land. Secondly, it established the comprehensive programme of indicators to guide the development direction towards sustainability as a critical activity of measurement. Thirdly, the special ecological cell-community-district-city framework has been designed to provide residents basic service functions and public activity space required for life within a short period walk, known as the human-scale range of daily activities.

### 4.3.1 *Geographical Condition: The Newly Built Eco-City*

Prior to Tianjin being shortlisted for the SSTEC site, the Tianjin Municipality had already paid attention to environmental concerns for the city's future development—in the *Tianjin Municipal Master Plan (2005–2020)*, one of main strategic planning goals is to build Tianjin into a liveable city. Located about 40 km east of Tianjin and 150 km to Beijing, the SSTEC sits within the Tianjin Binhai New Area. The site itself is on a former wetland area that had been used for industrial purposes. It is seen as 'a symbolic antidote to these polluted lands that can represent a new solution to addressing the ills of industrialisation and urbanisation' (Keeton, 2011, p. 128) (Table 4.1).

The eco-city concept nonetheless remains largely open to interpretation in terms of definition and focus (Chang & Sheppard, 2013; Lehmann, 2011; Pow & Neo, 2010). For example, in definitional terms, Joss (2009) has identified three broad eco-city categories that encompass most current eco-city developments—'newly-built eco-cities, the retrofitting of existing urban environments, and the expansion of existing urban areas' (p. 241). To allow for urban habitation, the SSTEC had to be rehabilitated and decontaminated, in order to successfully reclaim the wasteland into economically productive and habitable land. From this perspective, the SSTEC has been regarded as the newly built eco-city in this research, that is, having been built from the ground up (See Fig. 4.2).

**Table 4.1** General information about the SSTEC

| | |
|---|---|
| Launch Year | 2007 |
| Total Area | 31.23 $km^2$ |
| Geographic Conditions | Located in the core area of Tianjin Binhai New area, which is 45 km from Tianjin's city centre and 40 km from Binhai International Airport |
| Planning Goals | To establish a replicable eco-city that is resource-saving, environmentally friendly, economically robust and socially harmonious |
| Industries | Cultural creation, environmental protection, high technology and green building |

*Source* Information is from SSTEC Website

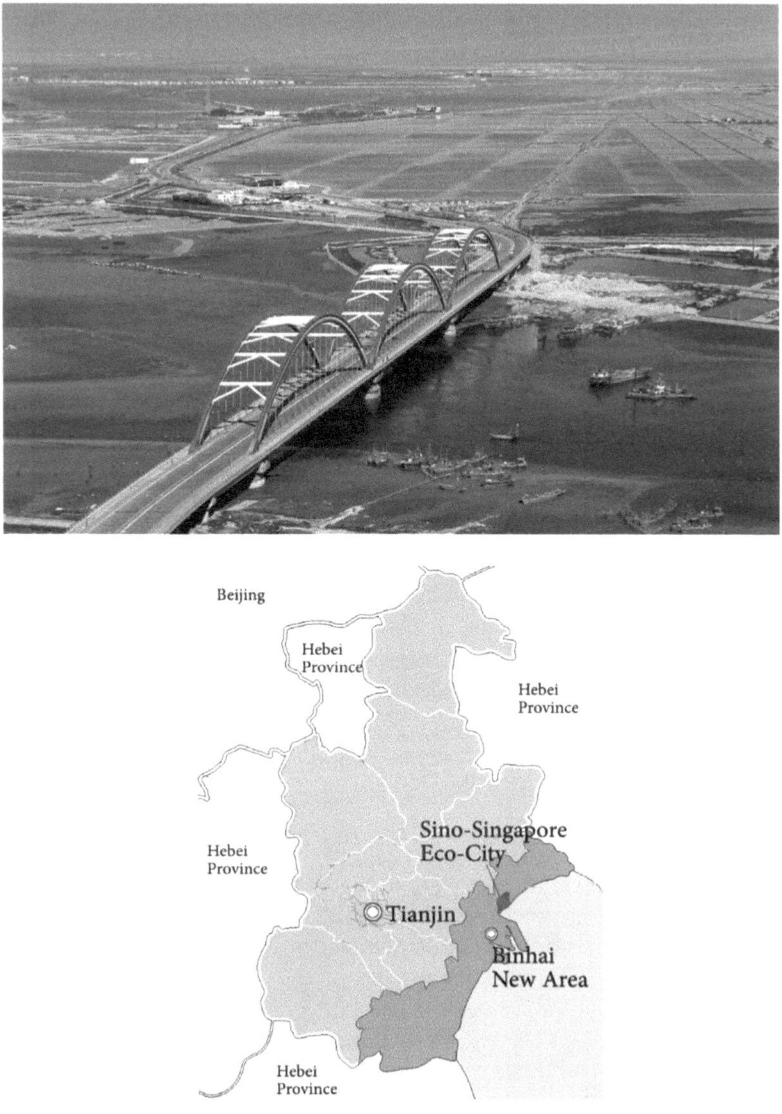


**Fig. 4.2** SSTEC before development and its location. *Source* Figure is from SSTEC Website. *Source* Map is from Xie et al. (2020, p. 10)

As being defined as the newly built eco-city, the SSTEC project is not the redevelopment project from any urban settlement as there are no inhabitants (people) at the beginning stage of construction. Because of its unique characteristics, on analysing this project without differentiating between stages, it would obviously miss the term of *people* as the core component of people-place relationship; as a result, it lacks the necessary elements in the overall discussion of people-place relationship in this research. By focusing on the post-occupancy (project operation) stage, the researcher can collect data in terms of residents' living experience in the eco-city, such as people's behaviours in public spaces and their attitudes towards the spatial quality of public spaces, and how is this eco-city project likely to be well managed and maintained to constantly keep the people-oriented planning goal.

Overall, the above-mentioned self-limiting characteristics of the SSTEC project have determined the procedure of data collection and further data analysis. Specifically, initial analysis about whether residents are satisfied or not towards the living experience can indirectly reflect whether the people-place relationship is harmonious or not in the SSTEC project. Then, data discussion about people–place relationship has been seen as the significant content basis for interviewing related project planners and managers. Such procedure of data collection and analysis are arranged to identify whether there are any evidences to show the global mainstream people-oriented planning concept has happened in the Chinese eco-city development, if yes, in what ways and to what extent.

### *4.3.2 Political Condition: Government-to-Government Project*

The SSTEC, being seen as a prominent national eco-city model, is the government-to-government flagship project collaboration between China and Singapore. It has a close relationship with the Chinese central government and the government of Singapore, together with the independent municipality of Tianjin. In November 2007, *The Framework Agreement between People's Republic of China and Republic of Singapore about Building an Eco-City in the People's Republic of China* was signed (Zhan et al., 2018, p. 5). It has been a highlight and a key project between these two countries, seeking for economically vibrant, environmentally friendly, resource-efficient and socially harmonious planning goals, in order to offer a reference for other Chinese eco-cities' planning and design about how to provide the people-oriented living settlement for living residents.

**Table 4.2** Explanation of the concept of *Three Harmonies*

| | |
|---|---|
| Social Harmony | People living in harmony with other people |
| Economic Vibrancy | People living in harmony with economic activities |
| Environmental Sustainability | People living in harmony with the environment |

*Source* Information is from the Singapore Ministry of National Development Website

**Table 4.3** Explanation of the concept of *Three Abilities*

| | |
|---|---|
| Practicable | The technologies adopted in the eco-city must be affordable and commercially viable |
| Replicable | The principles and models of the eco-city could be applied to other cities in China and even in other countries |
| Scalable | The principles and models could be adapted for another project or development of a different scale |

*Source* Information is from the Singapore Ministry of National Development Website

The SSTEC has been characterized as a thriving city that is socially harmonious, environmentally friendly and resource-efficient (Sino-Singapore Tianjin Eco-City Investment and Development Cp., Ltd. Website, no date shown), underpinned by the unique concept of *Three Harmonies* (三和), which could be interpreted as the traditional concept of sustainable development—the balance of social, economic and environmental concerns for both current and future generations. During the early planning process, both national governments had agreed for positioning the SSTEC to become as the norm-setting model, within a larger global context of variegated urban sustainability (Sino-Singapore Tianjin Eco-City Investment and Development Cp., Ltd. Website, no date shown), which are introduced as *Three Abilities* (三能)—practicable, replicable and scalable (Tables 4.2 and 4.3).

### *4.3.3 Bilateral Supervisory Management Framework*

Being recognized as the government-to-government flagship project collaborated between China and Singapore, the SSTEC project has been involved in the high degree of bilateral supervisory management framework. It was formalized in the *2007 Framework Agreement for the Development of the Sino-Singapore Tianjin Eco-City Project*, and based upon this agreement, the bilateral collaboration runs until 2020 when urban development is scheduled to reach the completion (Belle, 2015, p. 30). Due to this bilateral project feature, the management framework has been divided into two tiers—one focuses on more political supervisory committee, and the other focuses on operational and technical aspects.

Being co-chaired by the *Singapore Minister of National Development* and *Chinese Minister for Housing and Urban-Rural Development*, the *Joint Working Committee* oversees issues related to the masterplan, detailed plan and KPIs. It also oversees daily routines of the *Eco-City Administrative Committee* that is the local-level government administering to the SSTEC (Chen, 2016a, 2016b, p. 140). The *Eco-City Administrative Committee* itself comprises of different bureaus, such as 'environment bureau which is in charge of overall environmental management, the construction bureau which is in charge of drawing up the planning documents' (Ong, 2019, p. 58). At the building level, the *Eco-City Administrative Committee* is responsible for planning government-funded amenities like schools and hospitals, and government-funded infrastructure like roads, parks and transportations (Fig. 4.3).

### *4.3.4 Singaporean Government or Planning Authorities Involved*

On the one hand, at the Government-to-Government level, Singapore and China have committed to sharing expertise and experience in various aspects, such as urban planning, environmental protection, resource conservation, water and waste management, as well as in policies and programmes to engender social harmony in the eco-city. Specially, A *Joint Steering Council* co-chaired at the *Deputy Prime Minister-Vice Premier* level charts the strategic directions of the project. A *Joint Working Committee*, co-chaired by Singapore's *Minister for National Development* and China's *Minister of Housing and Urban-Rural Development*, and comprising senior representatives from agencies involved in the eco-city project, oversees the implementation of project and the achievement of its key milestones. On the other hand, at the private sector-level, the *Sino-Singapore Tianjin Eco-City Investment and Development Co., Ltd.* is formed by a Singapore Consortium and a Chinese Consortium, each with a 50% stake in the joint venture company. It is the master developer for this eco-city project. The company also takes on economic promotion of the eco-city. The undertaking of project by the private sector helps to ensure its commercial viability and enhance its scope for replication. *The*

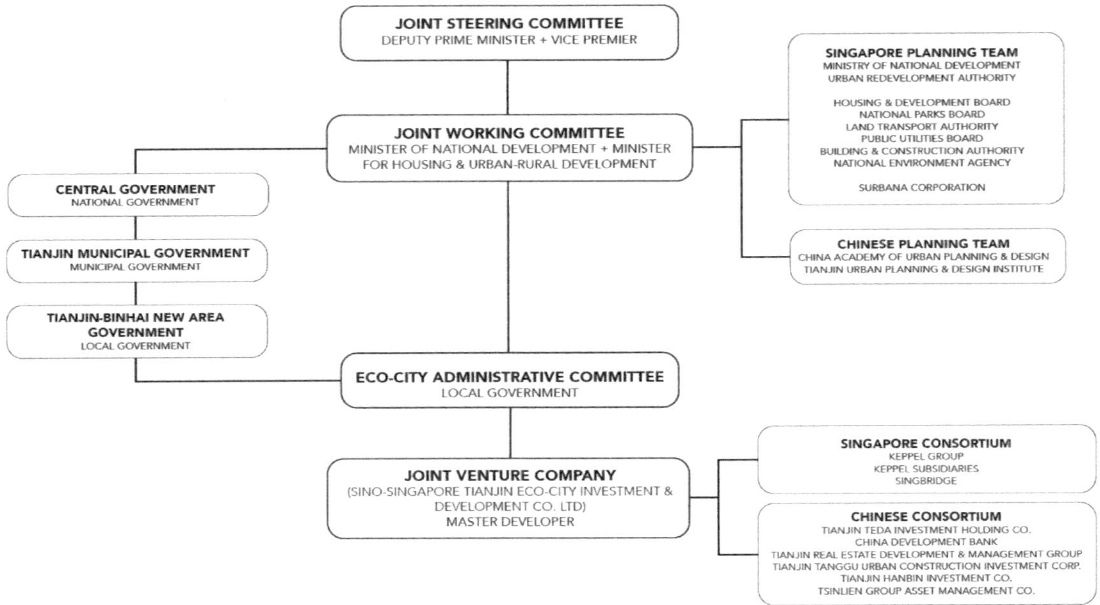


**Fig. 4.3** Bilateral supervisory and operational structure of SSTEC. *Source* Image is from Ong (2019, p. 57)

**Table 4.4** Memorandum of Understanding (MOU)

| |
|---|
| Amtronics Pte Ltd.—Green Building Research Institute |
| Formation of China-Singapore Green Construction Development Alliance to provide a platform for cooperation between 23 China and Singapore enterprises in the field of green construction, to jointly promote and exchange expertise in green and sustainable development of cities in China and along the One Belt and One Road |
| Keppel Land—Tianjin Eco-City Administrative Committee |
| Collaboration to enhance smart operations and municipal management services in the Sino-Singapore Tianjin Eco-City |
| PSA International—Tianjin Port (Group) Co., Ltd |
| Strategic partnership on smart port development, supply chain logistics and cooperation along the One Belt and One Road |
| Temasek Polytechnic—National Animation Industrial Park |
| Agreement to explore collaborations in internships placement; students and staff exchanges; access to training and consultancy expertise in the areas of innovation and entrepreneurship; and joint industry projects in Information Technology and serious gaming |

*Keppel Group* leads the Singapore Consortium, while the Chinese Consortium is led by *Tianjin TEDA Investment Holding Co., Ltd* (Ministry of National Development, Singapore Government Agency Website, 2021).

On 1 December 2020, Singapore and Tianjin signed a *Memorandum of Understanding (MOU)* to enhance the cooperation under *Singapore-Tianjin Economic and Trade Council (STETC)* at the 9th STETC meeting. It refers renewal and enhancement of cooperation between Singapore and Tianjin shall focus on priority areas of mutual interests, including sustainable and smart development, trade and investment, in order to spur Beijing-Tianjin-Hebei regional development (Table 4.4; Fig. 4.4).

## 4.4 People-Oriented Planning Concept Reflected by Indicator System

### 4.4.1 *The Key Performance Indicator (KPI): Translating the People-Oriented Planning Concept on the Ground*

The indicator index contributes to quantifying the development goal of spatial planning and to providing the foundation for implementation at the operational stage of urban management. Even though differences in assessment can be expected given a variety of analytical perspectives,

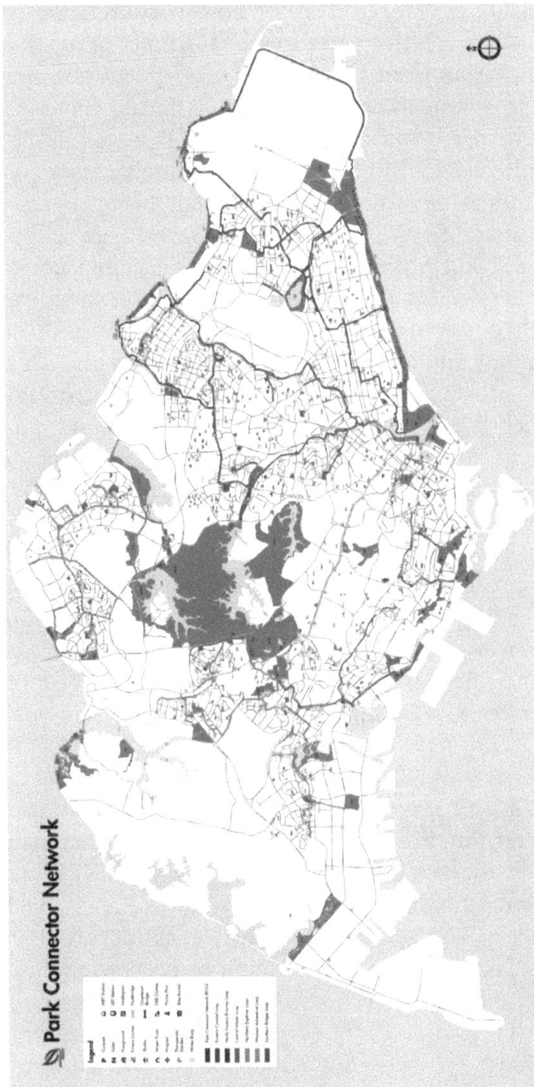


**Fig. 4.4** Park connector network in Singapore. 2022 *Source* Figure is from Cheng ()

debates about whether using indicators for measuring standard indeed emphasize the significance about the diverse analytical perspective of implementation gap. In the context of construction of Chinese new spatial planning system, it is necessary to seek the unity of rigidity and flexibility to adapt to the updated requirements for spatial planning reform. However, there is a fundamental need for Chinese eco-city planning to pay attention to persisting in the transition from economy-oriented indicator to people-oriented indicator. This research uses a comparative analysis of indicators to determine the implementation gap, especially for the humanistic aspect. As Wu (2023) demonstrated, in the process of community regeneration project, the number of facilities per capita was used as an assessment index in the past, although this index is intuitive, it ignores people's physiological needs and their opportunity selections. While the community model constructed based upon the walking-scale reflects the social attributes of indicator index. 'The streamlined indicator setting, which is related to development strategy and cuts in from people's perspective, strengthens the implementation of urban strategies and embodies the human nature of urban development' (Wu, 2023, p. 48).

The agreement of Sino-Singapore framework governs the operational mechanism, coordinated by a *Joint Steering Committee*. Referring to the Chinese aspect, leading institutions are the *Ministry of Housing, Urban and Rural Development*, others include the *National Development and Reform Commission*, the *Ministry of Finance*, the *Ministry of Environment* and *the Municipality of Tianjin*. The *Sino-Singapore Tianjin Eco-City Investment and Development Co. Ltd.*, is the master developer of the eco-city (Geroe, 2017, p. 989). Based upon its overarching planning concept of *Three Harmonies and Three Abilities*, such government-to-government project feature determines the KPIs have to be jointly formulated across countries to define the eco-city standards. Under the KPIs framework, planning targets are set to advance breakthroughs in areas, such as land use, water and energy usage efficiency, reduction about carbon emissions.

Unlike typical Chinese urban development projects which usually start with a masterplan, one of the project's significant impacts has been seen as being 'pioneered the approach of development a KPI framework before a masterplan is drawn up' (Sino-Singapore Tianjin Eco-City Administrative Committee et al., 2021. p. 42). The KPI is a critical component as the SSTEC is the first eco-city development that uses a KPI framework to guide its planning process and urban management system (Deng &

Cheshmehzangi, 2018, p. 132). It has been functioned to measure the eco-city's progress in environmental, economic and social development, and to ensure the design is well executed as well as policy and guidance is further applied at the implementation stage.

As the eco-city has been seen as a relatively broad concept, when it comes to whether a particular city is eco-city or not and to determine the development goal of eco-city planning, the indicator index need to be constructed to measure the implementation. The indicator index measuring the eco-city development has been regarded as the comprehensive evaluation system formed by a variety of indicators that reflect the stipulations of both quantity and quality of eco-city from various perspectives (Sun, 2010, p. 39). Moreover, Xia et al. (2021) pointed out the significance of first-level of index system about measuring eco-city's planning objective, the core of eco-city development is to meet the diversity of people's needs to make residents have a high evaluation of their own abilities and welfare of the whole society. Therefore, the aim of adding people's livelihood welfare into the first-level index system is not only to investigate long-term impacts on environmental, social and economic development, but also to evaluate whether eco-city is people-oriented (p. 32) (Fig. 4.5).

According to *Report on the Implementation of Indicator System (2020)*, to better promote the implementation in phases and time series, the latest indicator system (2.0 version), putting into effect in 2020, has adopted the three-steps and closed-loop implementation mechanism which includes decomposing implementation, monitoring data and evaluating feedbacks. Referring to the evaluation procedure, the SSTEC entrusts *Tianjin Eco-City Green Building Research Institute* as the third-party organization to carry out an annual evaluation about the overall situation of implementation, such as analysing the condition of completion and the trend of compliance. Afterwards, the SSTEC reports the evaluation status to relevant government departments, such as *Ministry of Housing and Urban-Rural Development* as well as *Tianjin Eco-City Environment Bureau*, in a timely manner. Those responsible institutions then give feedbacks to improve the comprehensive control of the overall implementation process.

On the one hand, two aspects have reflected the people-oriented planning concept in the first version of indicator index of the SSTEC, 'by providing the guarantee for residents' healthy lifestyle' (Sun, 2010, p. 40), the indicator of residential areas within five-minutes walking distance to

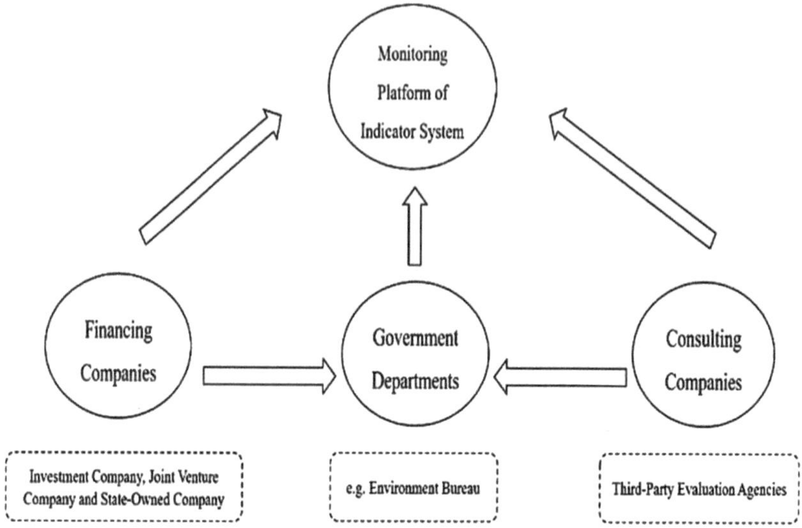


**Fig. 4.5** Departments involved in the implementation process. *Source* Figure is drawn by the researcher. 10 September 2021

parks and green spaces as well as the indicator of the ratio of residential areas within 500-metres accessibility of community service facilities. On the other hand, as the *Report of the Implementation of Indicator System of the Sino-Singapore Tianjin Eco-City (2020)* points out, indicators of two aspects, namely the domestic water consumption per capita per day and the coverage of barrier-free facilities in public space, have reflected the people-oriented concept (Table 4.5).

### 4.4.2 The Revised Indicator System (2.0 Version): An Ongoing Trend from 'Economy-Oriented' to 'People-Oriented' Development Pattern

Establishing the comprehensive programme of indicators has been regarded as a critical activity for measuring the distance towards, or away from sustainability. The most important function of indicators is to inform decision-making, which is about the reflection on relationship between planning policy and empirical data. As Bossel (1999) stated, 'indicators of sustainable development are needed to guide policies and decisions

**Table 4.5** People-oriented planning concept reflected by indicators

| *Focusing aspect* | *Report content* |
|---|---|
| The domestic water consumption per capita per day | In page 69, the indicator stipulates the domestic water consumption per capita per day in 2035 shall not exceed 110 litres per person per day. ... This indicator reflects the living standards of residents and the efficiency of water use, which not only *reflects the people-oriented concept*, but also avoids wasting and excessive consumption |
| The coverage of barrier-free facilities in public space | In page 133, the indicator stipulates that by 2023, the coverage of barrier-free facilities in public space shall reach 100%. ... Barrier-free facilities refer to the service facilities built in the construction project, to ensure the safe passage and the convenience of the elderly and children |

*Source* Information is summaries by the researcher

at all levels of society' (p. 4), helping policy-makers to improve their actions towards sustainability. Also, Joss et al. (2012) emphasized its significance, 'indicators constitute an institutional process of identifying policy, generating knowledge and applying that knowledge in practice' (p. 111).

Specifically, according to the *Report on Implementation of the Indicator System (2020)* jointly formulated by *Tianjin Eco-City Environment Bureau*, *Tianjin Eco-City Green Building Research Institute* and *Chinese Society for Urban Studies (CSUS)*, the framework of SSTEC's indicator system includes three levels, including the target level that determines the development vision of eco-city, the criterion level refined on the basis of the target level, and the indicator level refined on the basis of the criterion level. Specifically, the target level acts as a 'commander' by determining the top structure of SSTEC's indicator system. The criterion level regards the implementation way to realize the overall indicator target goal, and the indicator level guides the eco-city development by implementing detailed management indicators.

Compared with the criterion level of original indicator index (1.0 version), the revised one (2.0 version) has modified eight criterion levels and added three more levels to transfer its development path from economy-oriented to people-oriented. The number of total criterion levels has increased to fifteen, including good natural environment, liveable urban environment, healthy lifestyle, convenient social infrastructure, sound management, efficient resource utilization, technological innovation, city-industry coordination, etc. In terms of the 2.0 version, ecological environment and resource index has accounted for the largest proportion, reaching to 53%, social index accounts for 37% and the economic index declines to 10%. Moreover, in order to realize the transformation of development path from economy-oriented to people-oriented, urban management index and residents' experience at the operational stage are added to the type of social indicator index. Also, indicators of urban texture and liveable environment have been added to the ecological indicators in order to meet the green and sustainable development concept, indicators of urban texture and liveable environment have been added to the ecological indicators.

Regarding the public space aspect, the SSTEC has learnt from both domestic and overseas' experience on the basis of traditional Chinese planning indicators, to innovatively propose an eco-city indicator system and to establish the comprehensive evaluation system for eco-city's construction. As the *Report on Implementation of the Indicator System (2020)* shows, the monitoring method of the revised indicator system has taken residents' experience as a starting point and has assessed indicators by questionnaire surveys. Take the consideration of residents' experience as an example, the 2.0 version has increased the proportion of residential areas within a five-minute walk from parks and green space. And it also has increased the proportion of indicators that can directly reflect the improvement of people's sense of well-being, such as indicators of residents' community activities, their life satisfaction and health literacy level.

### *4.4.3 Driving Forces to Promote the Revision of Indicator Index*

Key Performance Indicators (KPIs) has been seen as an essential tool used widely to monitor the eco-city development (Science for Environment Policy, 2018). As Sharifi (2016) demonstrated, an important distinction between ecological urbanist movement and previous urban movements such as *Garden City* or *New Urbanism*, involves 'the extensive development of assessment toolkits used to verify and monitor project performance' (p. 8). These specific criteria for planning eco-city and monitoring its development effectiveness define a set of threshold or target indexes of environmental, economic and social standards. Changing policies and regulations caused by evolving development goals and planning priorities, the ever-increasing need of stakeholders' collaboration and the integration of ecology and intelligence driven by the technology have been three main drivers for promoting the revision of SSTEC's indicator index.

Before and during the conception of master plan, *The Ministry of Housing and Urban-Rural Development (MOHURD)*, which supervised the whole project, suggested to draw up a comprehensive system of performance indicators with the participation of *The Tianjin Academic of Environmental Sciences*. In 2008, *The Tianjin Academic of Environmental Sciences* therefore laid out environmental indicators by drawing on existing local and national indicators stored in the municipal records before submitting them to the *MOHURD*, which then added social and economic indicators. That is how the process of SSTEC's indicator index was produced and transferred to *The Administrative Committee of the SSTEC*. For the initial version of indicator index, publication of the twenty-six indicators has not been the end of story for this evaluation system. In fact, *The Administrative Committee of the SSTEC* decided to craft an operational breakdown of indicators in response to the demand from the *MOHURD*, which says 'KPIs must be implemented based on eco-city's design/construction/operational needs, also must be assessed and revised periodically improved' (Research Team of Tianjin Eco-City KPI, 2009).

The initial set of KPIs was developed in April 2008 and was based on the current Chinese national standards as well as best practices in Singapore, stating a number of these represent standards should be followed from the beginning, others should be realized by 2013, the remaining items should target by 2020. It is the fundamental reason for the revision of the indicator system due to the expiration of review time (See Table 4.6). Moreover, the influence of upper-level planning (上位规划) is another major reason for revision of the indicator system. Changes of national-level, regional-level or local-level planning policies can accordingly influence early-stage planning direction, medium-term planning assessment and later-stage planning implementation of sustainable projects. As discussed in Chapter 1, the Chinese existing urbanization model is seen as a product of traditional industrial era, while green urbanization is a leap from the traditional state to the new one. This change has given the breeding ground to the rapid development of Chinese eco-cities, or the so-called continuous driving force for eco-city's development. Under the changing influences of development mode, and also due to carbon reduction targets for the whole Chinese society in recent years, the revised indicator index introduces the latest environmental standards and incentivizes specific sustainable initiatives. In a manner of speaking, the revision of KPIs would be necessary to align with new planning policies and regulations under the development process of green urbanization, ensuring the SSTEC remains compliant and responsive to the changing legal framework to meet the development requirements in this new low-carbon era.

Throughout the implementation of SSTEC, stakeholders gather valuable insights and feedbacks based upon their experiences and observations. By analysing those lessoned learned, eco-city's project management team identifies areas that require improvement or adjustment. For instance, if certain KPIs do not effectively measure progress or if they do not capture specific environmental, economic or social concerns, indicators may be revised to better reflect the actual performance and address the identified issues. As is known, the SSTEC has followed a typical top-down method 'with a set of target indicators laid out by planners and sustainability specialists serving both Chinese and Singaporean governments' (Lin, 2018, p. 94). Influenced by economic interests and political agendas, the indicator index reflects objectives set by the administration, resulting in neglecting concerns of local communities. As Jin (2017) pointed out, 'initial making of the SSTEC and its governance involve little participation by residents for planning evaluation'. With improvement of urban management during the operational stage, the success of SSTEC relies on the involvement and collaboration of different stakeholders, including government agencies, private sector partners and especially local communities. To incorporate more stakeholders, the revised KPIs reflect a more inclusive set of metrics that address diverse interests and concerns of those stakeholders involved.

Technologies are celebrated as spectacles in pilot eco-city projects; this tendency is often reflected by KPIs, which are established to guide and to evaluate a city's development (Gudmundsson et al., 2016). Technological advancements contribute to playing a critical role in driving sustainability and offering new opportunities for enhancing the eco-city's

**Table 4.6** Time frame of KPIs of SSTEC

| *Category* | *KPI* | *Indicator value* | *Time frame* |
|---|---|---|---|
| Natural environment | Ambient air quality (days meeting National Ambient Air Quality II Standard) | > 310/Y | Immediate |
| | Quality of water bodies | Grade IV (China's standards) | 2020 |
| | Quality of water from taps | Potable | Immediate |
| | Noise pollution levels | Satisfy stipulated standards | Immediate |
| | Carbon emission per unit GDP | < 150 tonnes/$1 million | Immediate |
| | Net loss of natural wetlands | 0 | Immediate |
| Man-made environment | Proportion of green buildings | 100% | Immediate |
| | Native vegetation index | 70% | Immediate |
| | Per capita public green space | > 12 $m^2$/person | Immediate |
| Life style | Per capita daily water consumption | < 120 L/day | 2013 |
| | Per capita daily water generation | < 0.8 kg | 2013 |
| | Proportion of green trips | 90% | 2020 |
| | Overall recycling rate | 60% | 2013 |
| | Access to free recreational and sports amenities | < 500 m | 2013 |
| | Treatment of hazardous domestic water | 100% | Immediate |
| | Accessibility | 100% barrier-free | Immediate |
| | Services network coverage | 100% | 2013 |
| | Proportion of affordable housing | > 20% | 2013 |
| Developing dynamic and efficient energy | Usage of renewable energy | > 20% | 2020 |
| | Usage of water from non-traditional sources | > 50% | 2020 |

(continued)

**Table 4.6** (continued)

| *Category* | *KPI* | *Indicator value* | *Time frame* |
|---|---|---|---|
| | Proportion of R&D scientists and engineers in the eco-city workforce | > 50/10,000 workforce | 2020 |
| | Employment-housing equilibrium index (residents employed in the eco-city) | > 50% | 2013 |

*Source* Information is from Lin (2018, p. 96)

performance. For instance, advancements in renewable energy technologies, smart grids, building materials or transportation system like intelligent connected vehicle indeed improve energy efficiency, reduce carbon emissions and also enhance the overall resource management. The revision of KPIs would enable the eco-city project to incorporate these advancements, setting target and metrics that capture the adoption and effectiveness of these technologies. In the past decade, Chinese eco-cities are dedicated to explore how small and medium-sized emerging cities can embody the harmonious coexistence of people, economy and environment through smart development. In 2019, the SSTEC started to propose the development pattern of *All-Area Smart City (全域智慧城市)* guided by the smart city indicator index and following master plans with the *Strategy* of *Indicator-Leading and Planning-Control (指标引导规划管控战略)*. The SSTEC published the latest version of *Eco-City Smart City Master Plan (2021–2023)* in May 2021 to grasp the development process for future. According to *2020 Sino-Singapore Tianjin Eco-City Smart City White Paper*, in the next few years, the SSTEC aims to start building more intelligent and smart applications to provide solutions for the in-depth intersection of smart technology and residents' daily lives.

## 4.5 People-Oriented Concept Reflected by the Masterplan: To Provide the Liveable and Life-Convenient Environment for People

The Masterplan of SSTEC is prepared by multi-participative design teams of *China Academy of Urban Planning Design*, *Tianjin Urban Planning Design Institute* and *Urban Redevelopment Authority of Singapore*.

After the comprehensive discussion, the masterplan has been formulated, which is reviewed by *Tianjin Planning Commission* and related planning experts. Referring to the formulation process, based on the natural environment as well as the economic and social development level, the comprehensive survey has been conducted by means of planning. The masterplan aims to explore model of urban sustainable development and the ideal way of human settlement with six systems, including the ecological environment, the ecological economy, resource security, the ecological human settlement, the ecological culture and the ecological city management (Sun, 2010, p.38). The masterplan (2008–2020) firstly considers all requirements needed to achieve twenty-six indicators under the KPI framework, and then translates information into detailed plans, including land use, transport and energy use. It also conducts further research to support the detailed plan, 'covering area such as strategic positioning, population projection, circular economy, green transport system, green building, housing, and waste management' (Sino-Singapore Tianjin Eco-City Administrative Committee et al., 2021. p. 42).

The Masterplan of the SSTEC is followed by the traditional Chinese multi-planning integration working model—being based upon the higher-level planning, such as *Masterplan of Tianjin Municipality* and *Urban Planning of Binhai New Area* (Sun, 2010, p. 39). In the planning process of *the Masterplan of SSTEC*, both *Economic and Social Development Plan of the SSTEC* and *Ecological Environmental Protection Plan* were simultaneously compiled. *The Economic and Social Development Plan* implement requirements about the economic and social development to the urban space, by conducting an in-depth analysis of functions and city positioning of the SSTEC. *The Ecological Environmental Protection Plan* is not only an assessment of planning, but also this type of plan takes the environmental assessment as the starting point and development goal of planning, as it has been seen as the support of the masterplan by incorporating environmental requirements into eco-city's planning and strengthening the environmental management. In order to strengthen the implementation and management of eco-city's planning, the SSTEC has formed a spatial planning system with *the Economic and Social Development Plan* as the basis, *the Masterplan* as the support, *the Ecological Environmental Protection Plan* as the goal (Sun, 2010, p. 39).

The masterplan has been formulated to provide a clear blueprint to guide the SSTEC's overall urban development to achieve its KPIs, and to offer people a life-convenient living settlement under the general people-oriented design concept. Key special features have been seen as *One Axis, Three Centres and Five Districts* and *Four-Grade Public Service System*. One axis of the eco-valley spine is a slow mobility green network that goes through all centres and districts, aims to provide residents an easily accessible place for community interaction and outdoor activities to meet people's living needs for green open space. And the special ecological cell-community-district-city framework has been designed to provide residents basic service functions and public activity space required for life within a short period walk known as the human-scale range of daily activities.

The Masterplan (2008–2020) covers the initial 30 $km^2$ area for a population of approximately 350,000. The masterplan envisages its implementation from south-area to north-area of the site over three phases—'the first phase centres on the start-up area in the south of the SSTEC, the second phase focuses on the central district. And third phase focuses on the eco-island district, the northern and north-eastern districts' (Sino-Singapore Tianjin Eco-City Administrative Committee et al., 2021. p. 43). In the first phase, the masterplan aims to establish the basic spatial structure, including basic infrastructure, public facilities and transport network connecting it to the surrounding regions. Development plans in the start-up area around transport nodes and green spaces are of higher density (See Fig. 4.6).

The spatial layout of the SSTEC can be visualized as *One Axis, Three Centres and Five Districts* (See Fig. 4.7). The axis refers to the eco-valley spine, a slow mobility green network that goes through all centres and districts, linking the southern and northern surrounding areas. Three centres refer to the central business district and the two sub-centres both in the north and south. Five districts refer to the whole residential area. Also, the green and blue network emanates from the central green lung, Jing Lake and the park around it. It is worth mentioning the eco-valley intends to serve as the main public space in the SSTEC. It stretches 12 km through the eco-city in the north-south direction (as shown as the green line in Fig. 4.8) and incorporates water-sensitive urban design elements, such as eco-swales and dry streams. High-rise commercial residential buildings are built along the eco-valley. Cars are not allowed to drive through as it provides the green path for pedestrians and cyclists, aiming to build a conducive platform for community interaction and people's outdoor activities.

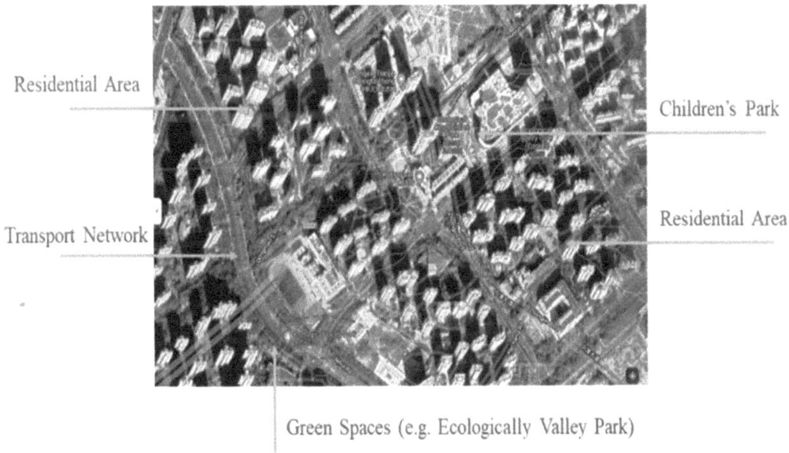


**Fig. 4.6** An example of residential area in the start-up stage. *Source* Figure is from Google Map (2022)

## 4.6 People-Oriented Concept Reflected by Spatial Indicators: Elements that Constitute Responsive Environment of the SSTEC

Public spaces suggest an image of accessible landscapes, specifically, the term public connotes the idea that landscapes are accessible to everyone, including people of a community, regardless of their age, gender or other characteristics (Altman & Zube, 1989, p. 1). In order to optimize the use of public spaces, Trancik (1986) stated the role of urban designer not merely to manipulate form to make spaces, but to create places through the total environment socially and economically (p. 114).

*Garden City* as a way to provide the classic theoretical foundation for the people–place relationship that supports eco-city's planning and design towards the humanistic direction. As demonstrated before, to make this conceptual relationship more concrete and much suitably applied into eco-city project, this research chooses public space as the representation of *place*. It aims to use the analytical framework of responsive environment to evaluate whether the environment of the SSTEC is 'responsive to the needs of people and is based upon the understanding that the relationship between the physical environment and social well beings of people' (Lyer,

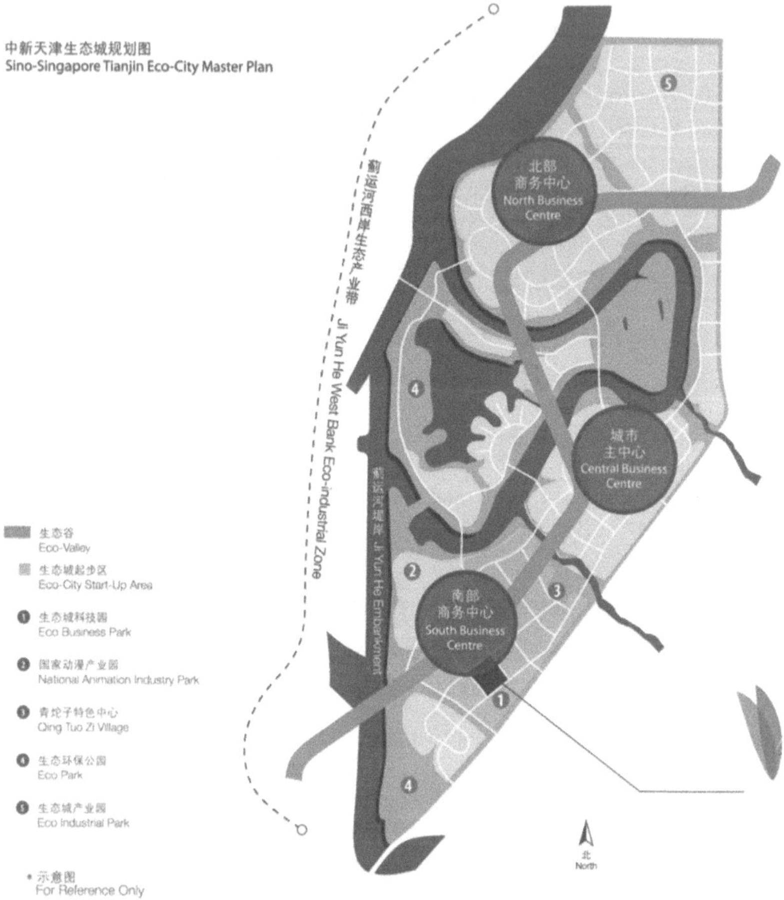


**Fig. 4.7** Eco-valley: the green path for pedestrians and cyclists. 2017 *Source* Image is from International New Town Institute Website ()

1995, p. 12). From this perspective, an essential part of understanding the responsive public space is to examine the performance criteria or qualities to identify appropriate indicators that can provide appropriate direction to an understanding of responsive environment. These performance criteria (See Table 4.7) have been detailed and extensively dealt with in the book *Responsive Environment: A Manual for Designers* (Bentley et al., 1985).

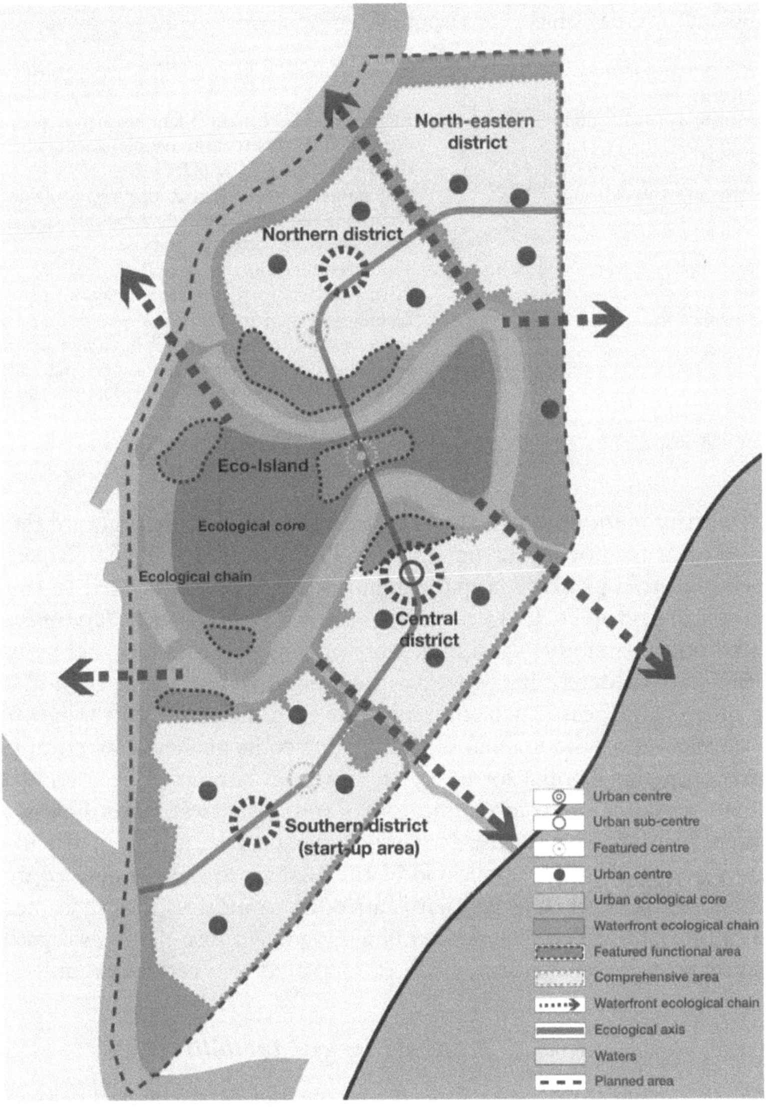


**Fig. 4.8** Masterplan of the SSTEC. *Source* Map is from Sino-Singapore Tianjin Eco-City Administrative Committee et al. (2021. p. 45)

**Table 4.7** Seven elements of responsive environment

| *Performance criteria* | *Meaning* |
|---|---|
| Permeability and legibility | Elements of access and identifies where people can go. It relates to how people orientate themselves within the place |
| Variety and robustness | The range and mix of uses that are available. Also, it is the degree to which people can use a given place for different purposes |
| Visual appropriateness and richness | The interpretation that people 'judge' the place. Also, it relates to the external image of place |
| Personalization | Elements related to contextual realities, and is concerned with detailed level of design where people take the initiative at the personal scale to personalize the place |

*Source* Information is from Bentley et al. (1985)

It is worth mentioning two issues concerning this text that require clarification prior to moving further analysis. On the one hand, the *Responsive Environment* is primarily a manual and is not explicitly related to sound theoretical principles. It is tied to the notion that *Responsive Environment* is a school of thought, thus, indicators of responsiveness are merely used as the tool to determine whether the design is responsive or not. On the other hand, reason for combining seven principles even though they are considered in the manual separately due to its analytical practicability. Specifically, this manual for designers has been chosen to be used as the primary resource to explore indicators of responsiveness, nevertheless, the content is located within the First World perspective, it is not particularly clear whether it is fully applicable to the analysis in the Chinese context. Therefore, the level of detail that is afforded to most of the performance indicators has been edited and combined, namely those elements regarded relevant to this research have been extrapolated to a certain extent.

### *4.6.1 Permeability and Legibility*

Only places which are accessible to people can offer them choice. The extent to which the environment allows people a choice of access through it from place to place is therefore a key measurement of its responsiveness (Bentley et al., 1985, p. 12). The greater the number of options given to users, the more permeable a place is. Also, Kevin Lynch (1984) regarded the access as a key performance dimension in order to achieve good city form, defining as 'the ability to reach other persons, activities, resources, services, information, or places including the quantity and diversity of the element which can be reached' (p. 118). Emerging from the definition of access is the relationship between people themselves as well as to place; hence, this relationship is responsiveness.

In order to determine whether the design is permeable and legible, a range of measurement tools can be identified. In order to promote permeability, it is for the place to be integrated at both regional and local levels. Referring to the large-scale level, a linkage to adjoining communities and major transportation routes needs to be taken into consideration, as the permeable access network can provide a choice of access to get from one place to another. While referring to the small-scale level, the public space through vehicular and pedestrian access would enable people to orientate themselves within the place, which also promotes legibility.

Permeability and legibility of the SSTEC case have related to prioritizing green travel, that is, planning the slow-traffic system within the eco-cell-sized neighbourhood to realize the separation of vehicles and pedestrians. As the green transport is a key feature of the SSTEC by increasing trips via public transport and non-motorized modes of transport within the eco-city. 'The SSTEC's road network deviates from Chinese traditional four-tiered road hierarchy. Instead, roads are divided into two categories, namely the two-way, six-lane major arterial roads and the two-way, four-lane secondary arterial roads' (Sino-Singapore Tianjin Eco-City Administrative Committee et al., 2021. p. 47). To encourage the active mobility, the SSTEC has implemented an extensive cycling network along the eco-cells which has separated pedestrian from vehicular traffic. As Sun (2010) pointed out, the slow-traffic system has the positive spatial nature and is closely related to people's activities, that is, it is a place for people to communicate and move. The design of the slow-traffic system pays more attention to the people-oriented design pattern. 'It is an open and continuous dynamic system based on scale and feeling of people. ... Important transportation hubs, eco valley, landscape nodes and waterfront open spaces of the eco-city are connected, providing passers-by with a safe walking space, and also considering barrier-free facilities' (Li & Feng, 2012, p. 54).

### 4.6.2 *Variety and Robustness*

'Variety relates to the range of uses that a place offers to its users, and robustness affects the degree to which people can use a given place for different purposes' (Bentley et al., 1985, p. 10). Essentially the extent to which variety offers can impact the degree to which places can be used for a variety of purposes. As Bentley et al. (1985) indicated, variety of experience implies places with varied forms, uses and meanings.

> A place with varied uses and has varies buildings types, of varied forms. It attracts varied people, at varied times, for varied reasons. ... Because the different activities, form and people provide a rich perpetual mix, different users interpret the place in different ways, it takes on varied meanings. (Bentley et al., 1985, p. 10).

Variety of use is therefore the key to variety as a whole. The purpose of promoting variety is to increase choice. That is, real choice has depended on the level of variety that prevails (Bentley et al., 1985, p. 29). And complexity refers to the aspect of people's responses to the place. 'The introduction of many elements having noticeable differences from one another increases the perceived complexity' (Nasar, 1998, pp. 75–76).

'A characteristic of positive environment is the intensity and diversity of activity' (Dewar & Watson, 1990, p. 20). Successfully performing environment is necessarily complex; they contain a variety of overlapping conditions and activities. 'They provide opportunities for the spontaneous and unexpected to occur and they accept conflict' (*ibid*). As Dewar pointed out, integration and complexity in a design are two elements that allow for a range of activities and responses that occur within a similar place to promote people's experience. Integration relates to how well functions overlap and in doing so, positively reinforce each other to create complexity. Complexity is also a result of responses in built environment and through a change in the place brought about by the integration of activities.

The SSTEC has adopted an eco-cell concept, similar to Singapore's township planning, to ensure provision of public social amenities within walking distance of residential estates. Currently, there are four community centres, drawing reference from Singapore's integrated Community Clubs, to provide common spaces for residents and to promote community bonding. There are also residents' centres, schools, kindergartens, hospitals and public parks linked by long park connectors. Having access to adequate public amenities and daily services is a crucial factor in strengthening the social fabric of the eco-city. For this purpose, public amenities are conscientiously planned and distributed within the SSTEC to ensure the majority of residents can meet their daily needs within a 15-minute walk from their home, and also public amenities are built ahead of demand to attract potential residents and to cater to the diverse interests and needs of residents. From this point of view, it reflects the forward thinking of eco-city planning in the aspect of people-centred design concept. Variety and robustness have related to the four-grade community service system. The most initiative of the people-oriented concept has been embodied in the SSTEC's design framework called *Four-Grade Public Service System*, which includes a set of eco-elements called *eco-cell*, *eco-community*, *eco-district* and *eco-city*. The aim of such land-use design is to provide people a liveable and life-convenient living environment by reducing journeys to work and amenities.

Li and Feng (2012) discussed the application of Singapore's model in SSTEC's community planning; the obvious feature is about the hierarchical community structure (p. 52). By considering newly built characteristics and combining with the *Code for Planning and Design on Urban Residential Areas (2002 Version)*, urban planning of the SSTEC has made the configuration of public service facilities not only have the ecological characteristic, but also has made public spaces have the human-scale spatial environment (Li & Feng, 2012, p. 52). Specially, as Fig. 4.9 shows, the SSTEC project supports a 400 by 400 metre city block, which has been called as an eco-cell. Roads line the perimeter of these cells, and each cell has the *grassroots community centre* with a service radius of 200–300 metres. Four of these eco-cells aggregate to form an eco-community to accommodate approximately 30,000 residents. Each eco-community has a residential community centre with a service radius of 400–500 metres. Figure 4.10 gives an illustration about an eco-community (eco-neighbourhood) of the SSTEC project—within the eco-cell, types of land use include residential, commercial, educational, industrial and medical areas, as well as parks and green space. Four to five of these eco-communities form an eco-district, and finally four of these eco-districts form the eco-city. At the district level, two sub-centres and the city centre form the commercial core.

It is worth mentioning that eco-cells, however, are separated from each other by not very porous boundaries, it lines up with people-centred planning concept due to the convenience of people's daily life. As the report

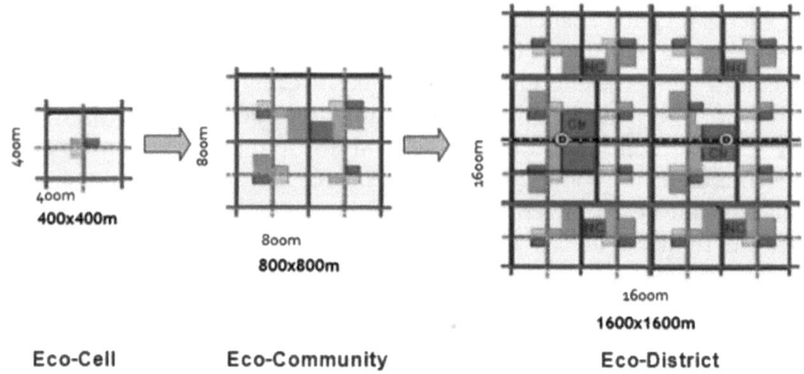


**Fig. 4.9** Unique land-use design of the SSTEC project. *Source* Image is from Sino-Singapore Tianjin Eco-City Investment and Development Co. Ltd Website. No Date Shown

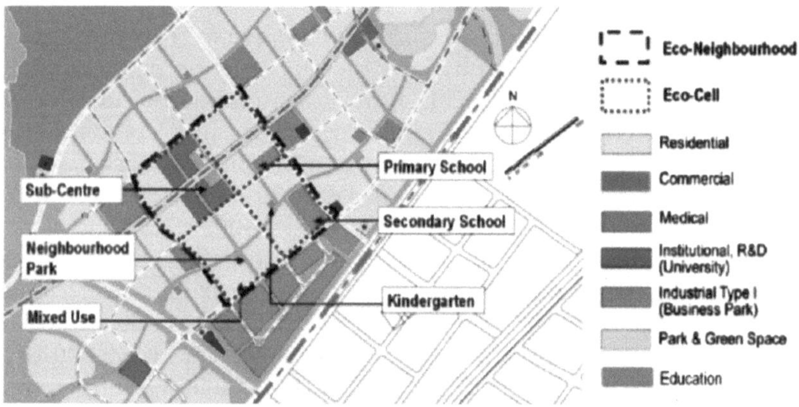


**Fig. 4.10** An illustration about the eco-neighbourhood. *Source* Image is from Chang (2017, p. 1726)

*Big Earth Data in Support of the Sustainable Development Goal (2020)* formulated by *China Academy of Science* points out, public space is a perquisite for improving the quality of life. Referring to SDG 11.7, it clearly proposes by 2030, China needs to provide convenient green public spaces within communities for all, especially for women, children, the elderly and those with disabilities (p. 76). In the post-pandemic era, public space is particularly significant for the sustainable development of communities, however, as people's preferences of life gradually become complex and diverse, the difficulty of community construction and management has increased accordingly. With the transformation about development pattern of Chinese urbanization, the spatial organization model has also changed from the previous residential space model of *residential area-residential quarter-residential group* to a hierarchical model of *15-min living circle* (Huang et al., 2020, p. 83). As the concept of *community life circle* (社区生活圈) has particularly received attention during the lockdown period, more thinking has been given to the functional connotation and construction mechanism of urban space. From this perspective, community construction with the concept of *community life circle* helps to think about eco-communities within the SSTEC. Even though they are large urban blocks of about 400 m by 400 m, and do not seem to be very porous, also, they are sometimes difficult to walk from one block to the other as they are separated, very often, by multi-lane highways. Design of eco-cells has been thought of as a human-scale proposition for how eco-city should be planned to improve people's quality of life—the planning goal is to provide people with all necessary destinations from their homes by walking or cycling within one block and within convenient commuting time. In turn, with more people interacting within local environment, the creation of vibrant neighbourhood and the promotion of place attachment are of great importance, which will be further discussed in the next chapter. Specifically, such spatial arrangement helps to promote green urbanization by reducing environmental pollution caused by excessive car use, lowing per capita energy consumption and reducing pressure on public health systems (Liu et al., 2020). Also, promoting walking in turn inspires the construction of parks, plazas and public spaces in communities, as well as improves the physical activity and mental health of residents (Zuniga-Teran et al., 2016). Meanwhile, for planners, by considering how to reduce commuting time to meet residents' living needs in their daily lives may contribute to rethinking the eco-city's transportation system to create more bicycles and sidewalks.

According to the *Technical Assistance (TA) Report (November 2009) Sino-Singapore Tianjin Eco-City: A Case Study of An Emerging Eco-City In China*, formulated in collaboration with the *Tianjin Municipal Government*, *Tianjin Eco-City Administrative Committee*, and the *Tianjin Eco-City Investment and Development Company Ltd.* and published by the World Bank, a hierarchy of live-work-play space refers to 'grassroots community centres, residential community centres, eco-city sub centres and the eco-city centre—all within the nested design framework' (p. 19). Such design framework of live-work-play space is equipped with both basic service functions and public activity space required for life within a short period walk (usually within a 15-minute walk), known as the human-scale range of daily activities. On the one hand, rather than promoting ad-hoc urban sprawl, the SSTEC area more intentionally adapts the urban structure to the requirements of pedestrians, cyclists and public transport. On the other hand, according to its masterplan, the SSTEC project proposes to provide the typical community where residents can walk 300–500 metres to a wide variety of destinations where people can enjoy leisure, recreational and cultural activities.

## 4.7 Current Development Trend of SSTEC

### 4.7.1 *Environmental Psychology Helps to Realize Development Goal*

By researching principles of environmental psychology, the development of SSTEC could be guided by a comprehensive understanding of users' needs and social well-being to achieve people-centred planning goal.

Firstly, environmental psychology mainly emphasizes on different groups of people as well as the degree to which the environment fits their needs. Even though disciplines of urban planning and environmental psychology are classified into the humanistic field, environmental psychology more closely relates to people and their needs concretely. Environmental psychologists make use of different research methods to gain a deep understanding of the target population's needs and preferences—they conduct surveys, interviews and focus groups to gather data on people's expectations, lifestyles choices and factors that influence their interactions with the built environment. Drawing on the urban planning field to guide the humanistic goal for eco-city development, environmental psychology can help to explore whether eco-city's residents prioritize proximity to green spaces for recreational activities, the availability of sustainable transportation, or the inclusion of community spaces. Such information could allow planners to cater to people-centred planning concept for design and amenities of eco-city's social facilities to meet specific requirements and desires of its residents.

Secondly, the focal point of environmental psychology mainly places particular emphasis on the micro level or on relatively small-scale environment, that is, environmental psychologists mainly focus on how the physical environment impacts social interactions. Drawing on urban planning field to guide the humanistic goal for eco-city development, it could be borrowed to apply to the aspect of community dynamics. By studying social behaviours in urban setting, environmental psychology can help to provide insights into design elements that promote social connections and the sense of community, such as assessing the impact of public spaces for people's gathering and mixed-use spaces on encouraging interactions within eco-city's residents. Then by incorporating these findings into eco-city's planning and design process, it contributes to creating an environment that fosters social engagement and strengthens community bonds.

Thirdly, environmental psychology mainly focuses on people's psychology and their surrounding environment, this theory believes different scales of the built environment have different impacts on human behaviours. Gibson (1986) claimed affordances do not change as people's needs change (p. 139), and the environment itself is meaningful and consists of opportunities for people's behaviours. On the one hand, environmental psychologists put emphasis on how the built environment impacts psychological well-being and health outcomes, from this perspective, by absorbing the environmental psychological thought, researchers in planning field can analyse social influences, motivations and barriers that have shaped residents' choices of energy use, waste management and green transportation. Specifically, by understanding the psychological drivers behind sustainable actions, planners could design interventions that encourage positive behaviour changes. And by addressing psychological barriers and leveraging motivations effectively, eco-city's development could promote sustainable lifestyles among its residents. For instance, urban planners could prioritize the inclusion of green spaces, promote human-scale walking and cycling paths, as well as design buildings that optimize natural lighting and ventilation. Such considerations contribute to creating healthy and conducive living environment within the eco-city. On the other hand, by absorbing the environmental psychological thought, researchers in planning field could conduct post-occupancy evaluation, surveys or behavioural observation to gather data on residents' psychological experiences and satisfaction level. Such feedbacks to some extent contribute to providing valuable insights into areas that require improvement and to helping planners to make behavioural data-driven decisions, in order to pay attention to people's needs and behaviours at the practical level.

### 4.7.2 From Technology-Driven Environmental Appeal to Environment-Technology-Humanity Tripartite Integration

The year of 2021 indicated the opening of *National 14$^{th}$ Five-Year Plan* and the beginning year of *Digital China Strategy (数字中国战略)*, making Chinese eco-cities step into a new era where ecology and intelligence become two main themes. Leading by the smart city indicators and the latest smart city master plan, the SSTEC has dedicated to create new leading experiences in the domains of digital economy, digital society, digital government or even digital ecosystem, fully unleashing the vitality of smart innovation.

In the past decade, an ever-increasing range of existing districts and cities, as well as new and planned urban initiative are labelled eco-districts or eco-cities under the carbon reduction target in China. However, the guiding planning documents tend to be largely developed as independent islands of locally ecological sustainability. Accordingly, it is more appropriate to think of eco-city as a planning approach that can be achieved through multiple ways to meet development requirements of green urbanization. Examples of eco-city models that emphasize technological solutions for achieving environmental objectives of sustainable development and green urbanization include *Symbio City* (Ranhagen & Groth, 2012), *Eco-Industrial Park* (Joss, 2011), *Carbon Neutral City*, *Zero Carbon or Low Carbon City*, *Eco2 City* (Suzuku et al., 2010) and *Smart Eco-City* (Bibri, 2020a, 2020b; Cowley, 2016).

The eco-city has been criticized as an idea that is 'loosely defined from a set of ostentatiously attractive projects' (Bibri, 2020a, 2020b, p. 9) as expensive schemes with aesthetic and commercial ends intend to satisfy local or regional ambition to invest in ecological sustainability without posing a more theoretically focused and global applicable approach. Cugurullo (2018) has questioned the sustainability of eco-city by investigating the extent to which it is developed in a controlled and systematic manner as its developers claim. Holmstedt et al. (2017) points out implementing sustainable solutions is more difficult as no unified practical definition is still accepted even if the subject of sustainability has been hotly debated more than three decades. In China, smart eco-city has become more popular and applicable to the current development appeal of green urbanization with technologies being used to deal with social infrastructures. As Bibri (2020a, 2020b) demonstrated, the eco-city is regarded as smart if there are 'intelligence functions that are able to integrate urban data to improve environmental sustainability through data-driven decision' (p. 11).

To mitigate the impact of COVID-19 on Chinese economy and also to meet *the 2030 Carbon Peak and 2060 Carbon Neutrality Goal*, in recent three years, the development vision of SSTEC is to create a holistic and exemplary city that balances economic growth, environmental protection, social well-being and technological innovation. Specifically, when SSTEC was firstly established in 2007, its primary objective has been seen as creating an environmentally sustainable city. As the original indicator index (1.0 version) shows, the SSTEC aims to minimize its ecological footprint by incorporating eco-friendly planning practices such as energy-efficient buildings, sustainable water management systems, waste reduction and recycling initiatives and the use of renewable energy sources. As this project progresses, the growing recognition of economic development and sustainability has been interconnected. The year of 2023 marks the 15th anniversary for the construction of SSTEC, its strategic focus has expanded to mainly foster economic growth and to attract investment, by means of transforming the SSTEC into a vibrant economic hub with a focus on knowledge-based animation industry. Planning initiatives including constructing technology parks and industrial zones aim to attract business and to create job opportunities. Simultaneously, with the advancement of technology, the SSTEC incorporates the planning concept of smart city into its strategic focus, and its development goal is to leverage advanced technologies and data analytics and to enhance eco-city's sustainability and residents' quality of lives. Planning initiatives include the deployment of smart grids for efficient energy management, intelligent transportation systems for optimized mobility, and the establishment of an integrated digital platform to provide smart services and to facilitate effective urban management. With the formulation of revised indicator system (2.0 version), there has been recently an increased emphasis on the social aspect of sustainable urban development within the SSTEC. Planning efforts have been made to improve public amenities, social infrastructures, healthcare services, education facilities and recreational spaces. Such a development goal is to ensure that the SSTEC is not only environmentally sustainable and economically vibrant, but also socially harmonious and being supportive of its residents' well-being.

## 4.8 Conclusion

The within-case analysis involves a number of basic elements of the SSTEC case, such as newly built eco-city due to its geographical characteristic, government-to-government project positioning and bilateral supervisory management framework due to its political characterize, which all provide essential foundation for the in-depth analysis in the following chapters. All content of this chapter aims to demonstrate the people-oriented concept is positive in both planning target and design initiatives.

The SSTEC has experienced a transformation in its strategic focus since its inception, that is, it has undergone three development processes which include initial focus on environmental sustainability, shifting towards economic development, integration of smart city technologies and liveability consideration. While environmental sustainability as well as the humanistic thought of harmonious coexistence between man and nature remain two key objectives, the SSTEC project has expanded to encompass planning strategies of digital economy and smart city technologies.

From the perspective of planning target, the aim of adding people's livelihood welfare into the indicator system is not only to investigate long-term impacts on environmental, social and economic development, but also to evaluate whether eco-city is people-oriented. Taking the consideration of residents' experience as an example, the 2.0 version of indicator system has increased the proportion of residential areas within a five-minute walk from parks and green space. It also has increased the proportion of indicators that can directly reflect the improvement of people's sense of well-being, such as indicators of residents' community activities, residents' life satisfaction and residents' health literacy level. All evidence-based discussion verifies that the indicator system of the SSTEC has continuously shown the ongoing trend of people-oriented development pattern.

From the perspective of design initiatives, the masterplan has been formulated to provide a clear blueprint to guide the SSTEC's overall urban development to achieve its KPIs, and to offer people a life-convenient living settlement under the general people-oriented design concept. Under the analytical framework of responsive environment to evaluate the SSTEC case, permeability and legibility have related to prioritizing green travel, that is, planning the slow-traffic system within the eco-cell-sized neighbourhood to realize the separation of vehicles and pedestrians and vehicles. Moreover, variety and robustness have related to the four-grade community service system. That is, the most initiative of the people-oriented concept has been embodied in design framework called *Four-Grade Public Service System*. The aim of such land-use design is to provide people a liveable and life-convenient living environment by reducing journeys to work and amenities.

## References


Adam, D., & Watkin, C. (2014). *The Value of Planning*. RTPI Research Report. [Online]. Available at: https://www.rtpi.org.uk/media/1548/value-of-planning-full-report-2014.pdf. Accessed on April 23, 2022.

Altman, I., & Zube, E. H. (1989). *Public Places and Spaces* (1st ed.). Plenium Press.

Asian Development Bank. (2010). *Eco-City Development: New and Sustainable Way Forward?* [Online]. Available at: https://www.adb.org/sites/default/files/publication/27868/urbandev-prc-nov2010-ecocity.pdf. Accessed on April 28, 2022.

Bentley, I., et al. (1985). *Responsive Environments: A Manual for Designers* (1st ed.). The Architectural Press.

Belle, I. (2015). *From Economic Zone to Eco-City – Urban Governance and Urban Development Trends in Tianjin's Coastal Area* (1st ed.). Borntraeger Science Publishers.

Bibri, S. E. (2020a). The Eco-City and Its Core Environmental Dimension of Sustainability: Green Energy Technologies and Their Integration with Data-Drive Smart Solutions. *Energy Informatics, 2020*(3), 1–26.

Bibri, S. E. (2020b). Smart Eco City Strategies and Solutions for Sustainability: The Cases of Royal Seaport, Stockholm and Western Harbour, Malmö, Sweden. *Urban Science, 11*(6), 1–42.

Bossel, H. (1999). *Indicators for Sustainable Development: Theory, Method, Applications. A Report to the Balaton Group* (1st ed.). The International Institute for Sustainable Development (IISD).

Chang, I.-C.C. (2017). Failure Matters: Reassembling Eco-Urbanism in a Globalizing China. *Environment and Planning A, 49*(8), 1719–1742.

Chang, I.-C.C., & Sheppard, E. (2013). China's Eco-Cities as Variegated Urban Sustainability: Dongtan Eco-City and Chongming Eco-Island. *Journal of Urban Technology, 20*(1), 57–75.

Chen, F. (2016a). The Deign Dimensions of China's Planning System: Urban Design for development Control. *International Planning Studies, 21*(1), 81–100.

Chen, G. (2016b). *Tianjin Eco-City: A Low-Carbon Model in Sino-Singapore Cooperation*. In *Singapore-China Relations: 50 Years* (pp. 131–152). World Scientific.

Cheng, I. (2022, February 22). *New Urbanism Perspective on Urban Planning of Singapore*. [Online]. Available at: https://urbandesignlab.in/new-urbanism-perspective-on-urban-planning-of-singapore/. Accessed on May 14, 2023.

Cowley, R. (2016). *Science Fiction and the Smart Eco-City*. The Society for the History of Technology Annual Meeting 2016. Singapore.

Cugurullo, F. (2018). Exposing Smart Cities and Eco-Cities: Frankenstein Urbanism and the Sustainability Challenges of the Experimental City. *Environment and Planning a: Economic Space, 50*(1), 73–92.

Dewar, D., & Watson, V. (1990). *Urban Markets: Developing Informal Retailing* (1st ed.). Routledge.

Deng, W., & Cheshmehzangi, A. (2018). *Eco-Development in China: Cities, Communities and Buildings* (1st ed.). Palgrave Macmillan.

Geroe, S. J. W. (2017). The Sino-Singapore Tianjin Eco-City: A Case Study of Chinese Experimental Regulatory and Institutional Development. *International Journal of Sustainable Development and Planning, 12*(6), 987–994.

Gibson, J. J. (1986). *The Ecological Approach to Visual Perception* (1st ed.). Lawrence Erlbaum Associates Inc.

Google Map. (2022, March 27). *Map of the SSTEC*. [Online]. Available at: https://www.google.co.uk/maps/place/Sinosingapore+Tianjin+ECO+City+Service+Centre/@39.1154217,117.7499684,3173m/data=!3m1!1e3!4m5!3m4!1s0x35f2782eb75eb85f:0xd57b86dcf3e668ae!8m2!3d39.1242241!4d117.743853. Accessed on March 27, 2022.

Gudmundsson, H., Hall, R. P., & Marsden, G. (2016). *Japan's Eco-Model City Program*. In *Sustainable Transportation: Indicators, Frameworks and Performance Management* (1st ed., pp. 277–295). Springer Berlin.

Holmstedt, L., Brandt, N., & Robert, K. H. (2017). Can Stockholm Royal Seaport be Part of the Puzzle towards Global Sustainability? From Local to Global Sustainability Using the Same Set of Criteria. *Journal of Cleaner Production, 140*(1), 72–80.

Huang, M., Wang, Y., Wang, C., & Zhang, J. (2020). Human-Oriented Life Circle Facilities Configuration ("生活圈"之辨——基于"以人为本"理念的生活圈设置配置探讨). *The Planner, 2020*(36), 79–85.

International New Town Institution. (2017). *Tianjin Eco-City, China, Asia*. [Online.] Available at: http://www.newtowninstitute.org/newtowndata/newtown.php?newtownId=1664. Accessed on March 27, 2022.

Jin, B. J. (2017). *Sino-Singapore Eco-City: Challenges Between A Ghost Town and A Sleeping Town (Chinese Version)*. Twenty-First Century Business Herald. [*Online.*] Available at: http://business.sohu.com/20130711/n381251056.shtml. Accessed on April 15, 2023.

Joss, S. (2009). Eco-cities: A Global Survey 2009. *WIT Transactions on Ecology and the Environment, 129*(May), 239–250.

Joss, S. (2011). Eco-Cities: The Mainstreaming of Urban Sustainability: Key Characteristics and Driving Factors. *International Journal of Sustainable Development Planning, 6*(2), 1–18.

Joss, S., Tomozeiu, D., & Cowley, R. (2012). Eco-City Indictors: Governance Challenges. *WIT Transactions on Ecology and the Environment, 155*(12), 109–120.

Keeton, R. (2011). *Rising in the East: Contemporary New Towns in Asia*. 1st ed. Amsterdam: SUN Architecture Publishers.

Lehmann, S. (2011). *The Principle of Green Urbanism: Transforming the City for Sustainability* (1st ed.). Earthscan.

Li, C. H., & Feng, C. Y. (2012). The Application of Community Model in the Residential Planning of the SSTEC. (邻里模式在中新天津生态城住区规划中的应用)*Cities, 2012*(3), 51–54.

Lin, Z. J. (2018). Ecological Urbanism in East Asia: A Comparative Assessment of Two Eco-Cities in Japan and China. *Landscape and Urban Planning, 179*, 90–102.

Liu, L. (2009). A Regional Divide in Sustainability Performance in China: Geography Governance and Income. *Sustainable Development, 19*(4), 250–262.

Liu, X., et al. (2020). Urban Form, Shrinking Cities and Residential Carbon Emissions: Evidence from Chinese City-Regions. *Applied Energy, 261*, 1–12.

Lu, D. D., & Chen, M. X. (2015). Several Viewpoints on the Background of Compiling the National New-Type Urbanization Planning (2014–2020). *Acta Geographica Sinica, 2015*(2), 179–185.

Lyer, N. K. (1995). *Towards Responsive Environments: A Case for Urban Design and Participation.*

Lynch, K. (1984). *Good City Form.* 1st ed. New York: The MIT Press.

Ministry of National Development, Singapore Government Agency Website. (2021, August 19). *Sino-Singapore Tianjin Eco-City: A Model for Sustainable Development.* [*Online.*] Available at: https://www.mnd.gov.sg/tianjinecocity/collaboration. Accessed on April 10, 2023.

Nasar, J. L. (1998). *The Evaluative Image of the City* (1st ed.). Sage Publications.

Ong, C. X. Y. (2019). *The Making of an Eco-City: An Examination of the Sino-Singapore Tianjin Eco-City as a New Model of Transnational New Town Development.* [Online]. Available at: https://dspace.mit.edu/handle/1721.1/123937. Accessed on March 27, 2022.

Pow, C. P., & Neo, H. (2010.) Building Ecotopia: Critical Reflections on Eco-City Development in China. In L. F. Lye & G. Chen (Eds.), *Towards a Liveable and Sustainable Environment: Eco-Cities in East Asia* (pp. 91–105). World Scientific Publishing.

Ranhagen, U., & Groth, K. (2012). *The SymbioCity Approach – A Conceptual Framework for Sustainable Urban Development*. SKL International.

Research Team of Tianjin Eco-City KPI. (2009). *Piloting the Eco-City: Implementing the KPI System for the SSTEC (导航生态城市:中新天津生态城指标体系实施模式)* (1st ed.). China Architecture and Building Press.

Science for Environment Policy. (2018). *In-Depth Report: Indicators for Sustainable Cities.* Produced for the European Commission DG Environment by the Science Communication Unit, UWE.

Sharifi, A. (2016). From Garden City to Eco-Urbanism: The Quest for Sustainable Neighbourhood. *Sustainable Cities and Society, 20*, 1–16.

Sino-Singapore Tianjin Eco-City Administrative Committee, China Eco-City Academy and Centre for Liveable Cities. (2021). *A Partnership for a Smart and Sustainable Future: The Sino-Singapore Tianjin Eco-City*. [Online]. Available at: https://www.clc.gov.sg//docs/default-source/books/a-partnership-for-a-smart-and-sustainable-future-the-china-singapore-tianjin-eco-city.pdf. Accessed on January 11, 2022.

SSTEC Investment and Development Co., Ltd Website. (No date shown). *Key Principles and KPIs.* [Online.] Available at: http://www.tianjineco-city.com/static/web/en/singapore3.html?lang=english. Accessed on March 18, 2019.

Su, H. J., & Wei, H. K. (2018). The Course, Inspiration and Prospects of China's Urbanization over the Past 40 Years of Reform and Opening-Up. *Reform, 2018*(11), 49–59.

Sun, X. F. (2010). The Preliminary Study on Eco-City Planning: Taking the Masterplan of Sino-Singapore Tianjin Eco-City as an Example (生态城市规划初探——以中新天津生态城总体规划为例). *New Energy and Green Building, 38*(8), 36–40.

Suzuku, H., et al. (2010). *Eco2 Cities: Ecological Cities as Economic Cities.* The World Bank.

Trancik, R. (1986). Finding Lost Space: Theory of Urban Design. *Van Nostrand Reinhold, 74*, 32–38.

Wei, H. K. (2014). *A New Road to Urbanization with Chinese Characteristics (走中国特色的新型城镇化道路)* (1st ed.). Social Sciences Academic Press.

Wu, F. L. (2007). Re-Orientation of the City Plan: Strategic Planning and Design Competition in China. *Geoforum, 38*(2), 379–392.

Wu, F. L., Zhang, F. Z., & Wang, Z. (2015). *Planning China's Future: How Planners Contribute to Growth and Development.* RTPI Research Report.

[Online]. Available at: https://fulongwu.files.wordpress.com/2016/02/rtpi-research-report-12-planning-chinas-future-december-2015.pdf. Accessed on April 24, 2022.

Wu, Z. Q. (2023). *Principles of Territorial Spatial Planning (国土空间规划原理)* (1st ed.). Press of Tongji University.

Xia, Y., Zhang, C. Y., & Zhu, H. (2021). The Performance Evaluation and Difficulties of the Eco-City Construction: The Case Study of Sino-Singapore Tianjin Eco-City. (生态城建设的绩效评价及难点探源——基于中新天津生态城的案例)*Cities, 27*(14), 27–40.

Xie, L. J., et al. (2020). Urban Entrepreneurialism and Sustainable Development: A Comparative Analysis of Chinese Eco-Developments. *Journal of Urban Technology, 27*(1), 3–26.

Yu, L. (2014). Low Carbon Eco-City: New Approach for Chinese Urbanisation. *Habitat International, 44*(2014), 102–110.

Zhan, C. J., Jong, M. D., & Bruijn, H. D. (2018). Funding Sustainable Cities: A Comparative Study of Sino-Singapore Tianjin Eco-City and Shenzhen International Low-Carbon City. *Sustainability, 10*(4256), 1–15.

Zuniga-Teran, A. A., et al. (2016). Designing Healthy Communities: A Walkability Analysis of LEED-ND. *Frontiers of Architectural Research, 5*(4), 433–452.

CHAPTER 5

# The Relationship Between People and Place from the Perspective of People's Experience

## 5.1 Evaluation of Spatial Quality and Significance of Interview Quotes

Evaluation of spatial quality for the selected research locations is based on two categories—people's activities (people) and indicators of the responsive environment (place). Referring to people's activities, the evaluation criteria draw from Gehl's approach, which is used to explore findings about people's low average staying time but have good spatial quality. Additionally, Masso's conceptual framework is used to assess how public spaces facilitate flows (movement and activities) and fixity (comfort and sense of place), striking a balance between these seemingly conflicting aspects to foster the place attachment in a highly mobile world. Moreover, this research uses the analytical framework of responsive environment to evaluate whether eco-city's environment is responsive to people's needs. It is grounded in the notion that the interactive relationship between the physical environment and people's social well-being is of great significance.

Based on outcomes of the observational analysis, this research suggests people develop the *behavioural connection* with the environment, and such connection reflects people's attitudes towards the place. To strengthen this argument, unpacking of interview excerpts serve as the methodological linchpin in exploring how people develop their



Y. Wu, *Possibilities and Limitations of Chinese Eco-City Development*, New Perspectives on Chinese Politics and Society,
https://doi.org/10.1007/978-981-95-1958-3_5

*behavioural connection* with the environment. By enabling residents to synthesize their viewpoints, these selected quotations foster a nuanced understanding about people's interaction with public spaces—in the good-spatial quality place, people are willing to spend more time for staying and interacting with others, while in the poor-spatial quality place, people tend to leave as long as they complete their necessary daily activities to express their dissatisfaction towards places.

## 5.2 Reasons for Choosing Four Main Research Sites and One Additional Research Site Within the SSTEC

This case study is regarded as being appropriate because it follows not only the consideration about being a newly built eco-city project with no past, history or time, but also and most importantly, for being famous for some of its achievements, such as being an environmental protection and socially harmonious planning approach, being a model of sustainable development, and seeing humanistic approach as its primary design principle. Therefore, understanding project's design concept and finding out its limitations contribute to making new insights about the changing public life in public spaces, making the eco-city development to be more sustainable and people-oriented in order to achieve the harmonious people–place relationship.

During the pilot study conducted in December 2018, in order to get better participatory experience, the researcher lived in the SSTEC for a month to experience the daily life just like an ordinary resident, and to explore 'what do people do and why do they do it, … [as well as to understand] the hidden meanings of a phenomenon' (Najafi et al., 2016, p. 1). During the main fieldwork period, the researcher already had a preliminary familiarity with the whole research site, noticing that the child prefers the children's park to play, the adult prefers the eco-valley to do exercise and the elderly prefers the community centre to get groceries and to gather together. On this basis, site selection focuses mainly on locations along two major axes of public life and leisure, with considerable criss-crossing of necessary, optional and social activities, a variety of public–private interface spaces—Zhongxin Avenue, the 3rd community centre, eco-valley and the children's park are chosen as four research sites within the SSTEC. For understanding whether the SSTEC is still an eco-city or it is now more of a smart and/or digital city in focus, during the supplementary fieldwork period, the researcher revisited the main four research sites and added the 4th community centre as the additional research site. Detailed reasons are explained as follows.

Firstly, since the aim of this fieldwork is to explore public life in public space within the SSTEC, the whole research area has been confined only to the public space, which can provide a chance for people to be together, to chat with each other, to walk or exercise and to enjoy themselves. Based on this selection criteria, public spaces have been chosen such as street, community centre and park, where full diversity of daily social lives are concentrated, and also where the main public and semi-public spaces, retail, dining entertainment are located. Secondly, in order to ensure all age groups are covered in the observation, the selection of research sites needs to cover different kinds of public space that are frequently used by each age group. That is, age group to some extent has determined the research site to ensure the diversity and objectivity of the research population. Also, public spaces where the crowds gather provide a convenient place for participatory observation during the research recording period; namely it is convenient to record outdoor activities for different ages of people. The general description about each research site has been shown as follows, including their geographical locations, functions and site characteristics. Also, the map of four study sites in the context of eco-city plan is shown in Fig. 5.1.

(1) **Zhongxin Avenue**

It is the main avenue when people enter the SSTEC. It always has a relatively heavy traffic and pedestrian flow. A few years after the eco-city was built, with the number of residents increasing, existing sidewalks on both sides of the avenue are too narrow to meet the needs of pedestrians. In 2018, the sidewalk along Zhongxin Avenue was widened to two metres, and the existing motor vehicle lanes remained unchanged.

(2) **The 3rd Community Centre**

The 3rd community centre—the first community centre constructed in the eco-city was officially put to use in 2013. It covers an area of 1.5

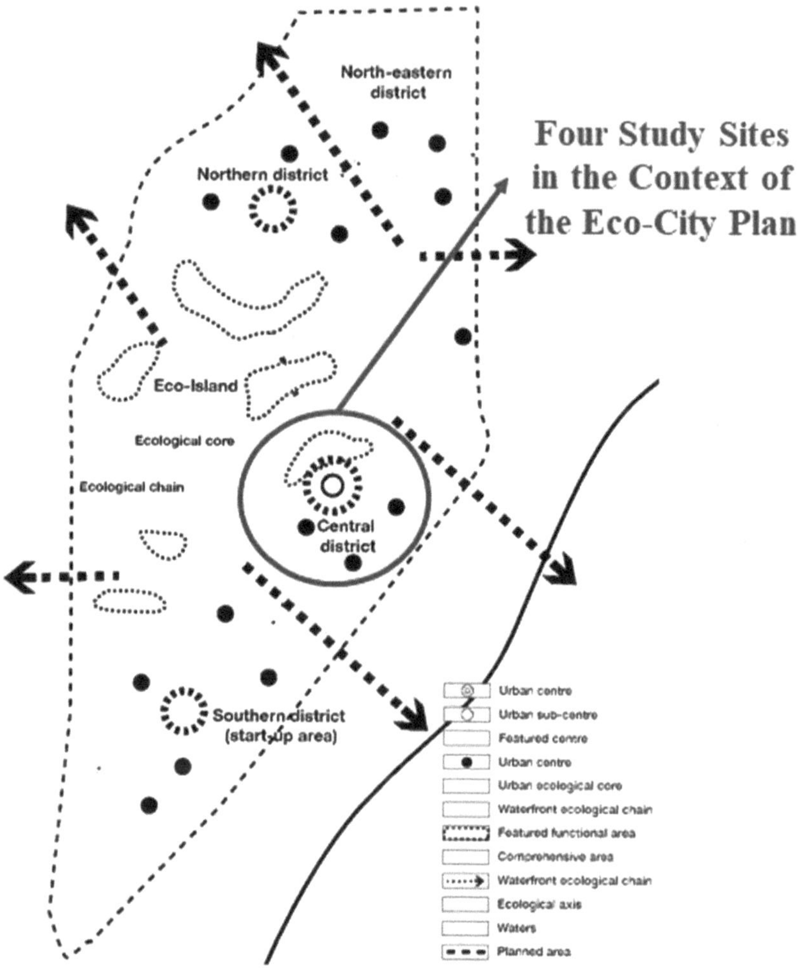


**Fig. 5.1** The map of four study sites in the context of the eco-city (*Source* Figures are drawn by the researcher)

hectares, and the commercial area is about 15,000 $m^2$. The indoor market was the first large-scale and eco-type food market for people's daily

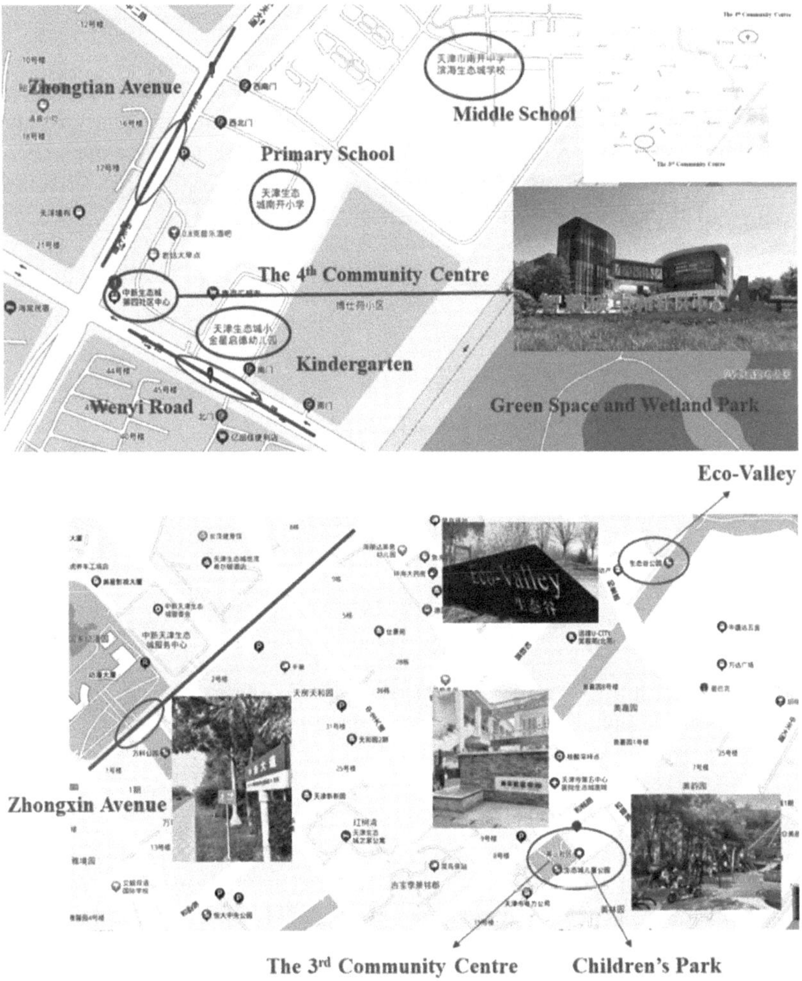


Fig. 5.1 (continued)

grocery shopping. The entire community centre aims to create a 15-minute life serve circle, including community management, health and commercial service.

(3) **Eco-Valley**

The 11 km eco-valley aims to connect residential districts, commercial zones and waterside areas as a green spine, providing connectivity and recreation space for 350,000 residents. Construction started in the year of 2008. This project demonstrates how bilateral cooperation can facilitate policy diffusion and project implementation in support of the large-scale development of green infrastructures.

(4) **Children's Park**

The children's park, the earliest community-level park, is adjacent to the 3rd community centre. To meet needs of different ages of children, this park has built various recreational facilities, including activity space for toddlers and teenagers, adventure park and other interesting activity venues. At nightfall, this park could be seen as the liveliest places for children in the eco-city.

(5) **The 4th Community Centre**

The 4th community centre is located at the intersection of Zhongtian Avenue and Huayi Road. The total built area is approximately 25,000 square metres. This newly built community centre is adjacent to one middle school, one primary school, one kindergarten, green space and wetland park. In terms of urban planning, it breaks away from traditional spatial barriers through an underground vehicle transportation system, and creates open and shard public spaces that is well-organized. Within the community centre, the commercial centre is developed by *Tianjin Eco-City Investment and Development Co., Ltd* and operated by *Tianjin Eco-City Urban Resources Operation Co., Ltd*, emphasizing low-carbon and smart characteristics.

## 5.3 Research Data Collection Time (PSPL Survey)

Being the observer, the researcher took notes and mapped people's activity during the participant observation period, by using the pre-prepared site map and template to record data. Moreover, the fixed time slot has been adopted in order not to presuppose people's activities and

**Table 5.1** Time horizon and condition

| *Time condition* | *Time horizon* |
|---|---|
| Seasons when the weather is suitable for travel (sunny and cloudy days) | Winter (December 2018)<br>Summer (July 2019 and June 2023) |
| Time covers working days and weekends | At least two recording times for each working day and weekend in each research site |
| 'Time-out' hours | Weekdays: lunchtime activity peak (1pm–3pm) and after working hours (6pm–8pm);<br>Weekends: (10am–8pm) when retail activities are open |
| Peak-hours | Mornings (8am–10am) and Sunset time (6pm–8pm): more people taking buses going to or coming from work, and doing exercise after all-day work |

*Source* Information is summarized by the researcher

patterns for using social infrastructures. Time-frame defines certain hours of the day, consisting of observations in the morning (8am–10am) when people go to work at the peak hour, at noon (1pm–3pm) when people have lunch outside and in the evening (6pm–8pm) when people go home after work and take outdoor exercise. Two days were spent per research site, one working day and one weekend respectively, during a total six-weeks period (three weeks in winter and three weeks in summer) with the good 'time-out' conditions for participant observation. In short, this time strategy allows a good cross comparison about people's outdoor activities in different seasons, which can help to give the researcher a real picture of public lives within the eco-city (Table 5.1).

## 5.4 The Evaluation of Spatial Quality of Each Research Location

Public space has been studied widely in literatures, including its visual features (Lynch, 1960) and aesthetics (Cullen, 1961). Its design features help to prevent crime (Newman, 1973), relationship between people's behaviours and spatial quality (Gehl, 1987) and evaluation about spatial quality of public space (Cooper & Francis, 1998). More recently, the restorative impact of green space on users has been widely researched

(Aspinall et al., 2015; Catharine, 2011; Hartig, 2007). This research more focused on Gehl's interests as his research mainly discusses the public space and public life, which responds to the core theme of this research, namely people-oriented eco-city development should consider people's needs and behaviours.

As this research aims to examine the particular people–place relationship in Chinese eco-city project, general classification of spatial quality for the selected four research locations are evaluated by types of people's activities and by some indicators of the responsive environment. Detailed discussion about the evaluation has been shown as follows.

### *5.4.1 Evaluation by People's Activities (People)*

As the research data about classification of people's activities in each site (*See* Appendix E) shows, the common characteristic of public life in the SSTEC is the versatility and complexity of outdoor activities, with some overlapping and shifts among walking, resting, playing, shopping and so on. As Gehl (2010) described, 'unpredictability and unplanned, spontaneous actions are very much part of what makes moving and staying in city space such a special attraction' (p. 20). In the public space, a clear core pattern has emerged from the great diversity of people's activities. One simple way to analyse the research data about people's activities is to put the most significant categories on a scale based on their degree of necessity. At one end of the scale are the purposeful necessary activities, such as walking between the bus stop and the destination, waiting for bus in Zhongxin Avenue, which take place under different weather conditions, no matter rain or shine. At the other end of this scale are the largely optional and social activities that people prefer to do in the public space, such as standing and chatting as well as baby sitting and children playing in the 3rd community centre, sitting on benches and doing exercise in the eco-valley.

Gehl (2010) defined the connection between (people's) outdoor activities and (place's) outdoor quality in his book *Life Between Buildings*—an increase in spatial quality gives a boost to optional activities, and increase in activity level then invites a substantial increase in people's social activities (*See* Fig. 5.2). Specifically, in poor-quality places, only strictly necessary activities occur, while in good-quality places, necessary activities take place with about the same frequency, or even a longer time, due to the better built environment. More importantly, a wide range of optional activities will also take place as the pleasant built environment invite people to rest or to play. As Gehl (2010) presents, 'in streets and city spaces of poor quality, only the bare minimum of activity takes place. People hurry home. In a good environment, a completely different, broad spectrum of human activities is possible' (pp. 21–22).

Based on the relationship between outdoor quality and activities outlined above, research data from participatory observation is analyzed by compiling and cross-referencing three categories of data—types of people's activities, people's average staying time and pedestrian flow rate. Those four research locations have been divided into two parts—good-spatial quality and poor-spatial quality public spaces. Table 5.2 shows the specific reason about such division. It is worth mentioning even though the 3rd community centre has the high people flow rate and the majority of people's activities are optional and social activities, people tend to stay only for a short time. Combined with the site observation, such as lack of shade and resting area, the 3rd community centre is seen as the poor-spatial quality public space.

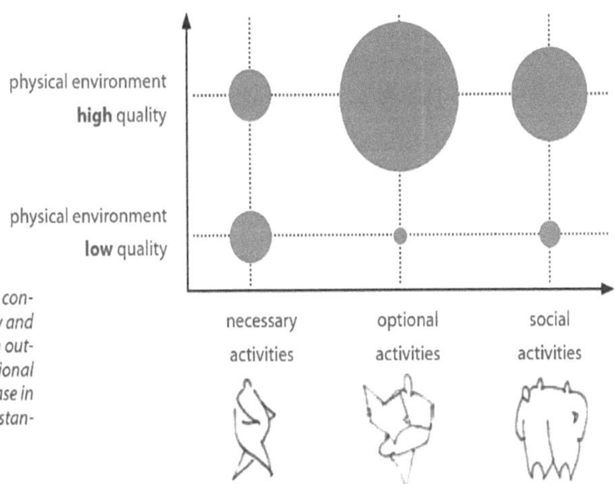


*Graphic representation of the connection between outdoor quality and outdoor activities. An increase in outdoor quality gives a boost to optional activities in particular. The increase in activity level then invites a substantial increase in social activities.*

**Fig. 5.2** Connection between outdoor quality and activities (*Source* Figure is from Gehl, 2010, p. 21)

**Table 5.2** Division about spatial quality by people's behaviours

| *Research location* | *People's behaviours* | *Spatial quality* |
|---|---|---|
| Zhongxin avenue | Mainly necessary activity<br>Low average staying time | Poor |
| The 3rd community centre | Mainly optional and social activities<br>High people flow rate<br>Low average staying time | Poor |
| Eco-valley | Mainly optional and social activities<br>High average staying time | Good |
| Children's park | Mainly optional and social activities<br>High average staying time | Good |

### *5.4.2 Evaluation by Indicators of Responsive Environment (Place)*

As discussed in the previous chapter, the concept of responsive environment is situated within the global mainstream's theoretical framework, particularly as represented by American and Northern European perspectives. However, it remains unclear whether this framework is fully applicable to analyse within the Chinese context. To maximally make this analytical tool practical, the evaluation of spatial quality for each research location has been focused on some particular indicators related to each location's spatial features, which is based upon different spatial characteristics/issues found out in the pilot study by researcher's observation experience—indicators of legibility and permeability are related to Zhongxin Avenue and the Eco-Valley, indicators of variety are related to the 3rd community centre and children's park (Table 5.3).

As Chapter 2 demonstrates, designing the liveable place for people deems as the core aim of eco-city planning, and creating the sense of place for people not only depends on the form of physical environment, but also relies on meaning and people's activities, other two equally important elements. In this research, people-oriented eco-city planning has been deemed as meeting people's needs, thus, in the following analysis, people's activities in public spaces and their perception towards public spaces are main points for discussion, while spatial elements tend to be causes of people's satisfaction or dissatisfaction.

**Table 5.3** Division of spatial quality by the responsive environment

| *Research location* | *Spatial features* | *Spatial quality* |
|---|---|---|
| Zhongxin avenue | (1) The street layout can offer a choice of direct routes to destinations (e.g. bus stops and residential areas), but too much road area has been provided for car use, which cannot make people feel comfortable walking experience due to narrow sidewalk, that is, it cannot be possible for pedestrians to move freely<br>(2) It is not structured to allow easy orientation and navigation for people, which is not user-friendly to the elderly and the blind | Poor legibility and permeability caused by not well-connected street layout |
| Eco-valley | (1) The permeable street layout offers a choice of direct routes to destinations (e.g. residential areas). And its layout can encourage people's comfortable experience of outdoor activities, such as walking, running and cycling<br>(2) It is structured to allow easy orientation and navigation for people, which is user-friendly to different age groups | Good legibility and permeability caused by the well-connected street layout |
| The 3rd community centre | (1) As the community centre aims to create a 15-minute life serve circle, people's daily routine to come here is mainly seen as grocery shopping and gather together for chatting<br>(2) This community centre aims to attract people's uses for different age groups, but the using group mainly focused on the elderly | Poor variety of uses compared to the planned spatial functions |
| Children's park | It is structured to allow easy orientation and navigation for people, which is user-friendly to different age groups | Good variety of uses compared to the planned spatial functions |

*Source* Table is drawn by the researcher

## 5.5 People Measure the Spatial Environment by Their Behaviours

As is shown in Chapter 2 about how the concept of planning for people reflects in the *Ecological Urbanism* theory, one of the theoretical contributors Kelvin Lynch (1981) suggested that criteria of good city form is about how well urban form sustains life (vitality), how clearly it is perceived in space and time (sense), how well environment and behaviour 'fit' and whether these standards are presented in a manner that provides 'access, control, efficiency and justice' (p. 11) and offers people the liveable and life-convenient environment. As one of the vital criteria of good city form, the 'fit' of environment and behaviour focuses on the discussion about relationship between the space and people's behaviours. This section plans to use the initial research data collected by the PSPL Survey to illustrate how people develop the behavioural connection with the space.

Based on outcomes of the observational analysis, it is the viewpoint of researcher that people develop the *behavioural connection* with the space, and this relationship indirectly reflects people's attitudes towards the place. Specifically, people's interaction with the (public) space could be seen as the result of their cognitive abilities to evaluate the spatial quality and to express their satisfaction/dissatisfaction about the (public) space by their behaviours—in the good-spatial quality place, people are willing to spend more time for staying and actively interacting with others. While in the poor-spatial quality place, people tend to leave as long as they complete their necessary daily activities in order to express their dissatisfaction about the place.

This section intends to explore how people perceive/measure place by their behaviours in public spaces within the eco-city. It is related to the research method of observation (mapping and on-site counting). Detailed discussion about how people interact with the spatial environment has been divided into two aspects—in both good-spatial and poor-spatial qualities, to explore how people develop their *behavioural connection* with the place.

### 5.5.1 *Good Spatial Environment and People's Behaviours*

On the one hand, as the community park is for people who live near Vanke Jinlu Community to do daily exercise and to relax themselves, the eco-valley has a relatively high average staying time during the whole recording period—apart from the selected summer weekend (13 July 2019, 8am–10am), people of different age groups all spend more than 30 mins staying in this public space in the two-hour flow record. Take the selected weekday (Monday, 15 July 2019) as an example. During the whole observation, at daytime, most of the space using group are the retired elderly (over 65 years old), children and women. The elderly people sit under the shade of trees and chat with their neighbours. Children play with their friends in this public space, such as making sand-castles, singing. The major group of caring for children is the middle-aged women. In order to easily look after their children, some of these middle-aged women stand near their children, and some of them sit on benches and chat with other neighbourhoods, however, they always keep their eyes on the children playing nearby. As the spatial analysis of eco-valley shows, this public space has good views that people can stare into the distance. Most of the space-using group at night time are young people. After a whole day's work, they do approximately one-hour fitness activities (e.g. walking and jogging) in the eco-valley to eliminate the fatigue of the day. It is obvious that a variety of outdoor activities occur in the same outdoor environment, but different user groups during the day and night result in different functions of this public space, making this place a multi-functional public space (Figs. 5.3, 5.4, and 5.5).

On the other hand, as the significant public space for children to play and to make friends, it is the earliest community-scale children's park putting into use in SSTEC. This park is adjacent to the 3rd community centre; it has been seen as the most lively 'happy valley' (defined by one of interviewees, female, about 35 years old) for different ages of children, especially at nightfall and at weekend. Some transcript of interviews to express their satisfaction has been shown as follows.

① *My grandson especially likes this place. It is summer vacation, he is eager to play with his friends here every morning, despite the high temperature. I often do shopping in the nearby supermarket (the 3rd community centre) and take him to play here about one hour, then go home. (Resident, female, about 65 years old) (Interview time: 10.20am, 16 July 2019).*

② *It is so interesting and I can enjoy myself here. My classmates and I often come here to play games after class. I can also make new friends here after playing with them. However, I think more sports facilities should be provided here, as our boys like playing basketball and riding skateboard. (Resident, girl, 10 years old) (Interview time: 1.30 pm, 14 July 2019).*

③ *My husband and I are used to come here after dinner. He plays with our granddaughter, and I participate the square dancing with my neighbours (middle-aged or elderly women). We go home together*

**Fig. 5.3** People exercise in the eco-valley

*before 9am, and the retired life is so happy. (Resident, female, about 60 years old) (Interview time: 6.15 pm, 16 July 2019).*

④ *Only the third community is equipped with the children's park, even some parents who live in the first and second communities often take their children to play here. There are quite a lot of people here during the peak hours (e.g. 9–10am and 6–8pm). But I am wondering why there are no children's park in the first and second communities? (Resident, female, about 35 years old) (Interview time: 10.40am, 16 July 2019).*

However, it has a very low flow rate during the whole recording period–in the selected summer weekday (16 July 2019, 1pm–3pm), the total number of people in this two-hour flow record is 31, while people's

**Fig. 5.4** People in the resting area in the eco-valley

average staying time is all over 30 mins. In the selected summer weekend (14 July 2019, 1 pm–3 pm), the total number of people in this two-hour flow record is only 13, however, people's average staying time is approximately 40 mins. It might because at the lunch time that people tend to stay in this public space for a longer time. Comparing with another time slot (16 July 2019, 8am–10am), the number of people for each type of activity does not exceed 30, however, people's average staying time is all above 30 mins. It is worth mentioning that the average staying time of children playing has reached 60 to 90 mins. By similarity contrast, the most obvious characteristic of this public life research data is seen as low people flow rate but high staying time. As the children's park, most of the space using groups are children aged 4–12 years old and their family members (e.g. the retired elderly and the middle-aged women). Research data from participant observation shows that during three different time recording period, children usually spend an hour playing with others in this park. They enjoy a variety of playground equipment, such as slides and swings (*See* Fig. 5.6). Meanwhile, adults who care for their children

**Fig. 5.5** Children playing in the eco-valley (*Source* Images are taken by the researcher. 13, 15, 19 July 2019)

sit in the resting area where tables and chairs are provided (*See* Fig. 5.7). They chat with each other, sharing their daily lives and experience of caring for children. It is noteworthy that adults' eyes are never away from their children who are playing nearby, such phenomenon shows place has good views that people can stare into distance.

### 5.5.2 *Poor Spatial Environment and People's Behaviours*

Referring to the poor spatial quality of public space, namely the Zhongxin Avenue and the 3rd community centre, research data of participatory observation is shown as below.

On the one hand, as the significant public space within the entrance area of SSTEC, Zhongxin avenue has a high flow rate during the whole recording period—in the selected winter weekday (20 December 2018, 8am–10am), the number of people in this two-hour flow record has

**Fig. 5.6** Children playing in the children's park (*Source* Figures are taken by the researcher. 16 July 2019)

**Fig. 5.7** Adults sitting in resting areas in the children's park (*Source* Figures are taken by the researcher. 16 July 2019)

reached 537. During the two-hours recording time, as it is the morning peak period, the majority of people are commuters, and some are parents who send their children to school. After 9 o'clock, the majority of people are the elderly who go out to shop for groceries. And in the selected winter weekend (22 December 2018, 1pm–3pm), the number of people in the two-hour flow record has reached 738—the majority of people are the young people, walking between bus stops and destinations as well as waiting bus (*See* Fig. 5.8), and some are teenagers who travel together. According to these research data, even though the major using group of those two days is different—the retired elderly people are the major group in weekday while the young people are the major group in weekend, people's average staying time is both approximately 15 mins, which means they stay in this place only for a short time. In addition, during the whole recording time, the researcher observed that most people hurry to their destination without any more outdoor activities, such as chatting and greeting; their fast-walking speed seems to tell others that they want to escape this place quickly.

On the other hand, as the significant public space for people to do shopping and other daily activities in SSTEC, the 3rd community centre has a relatively high flow rate during the whole recording period—in the selected summer weekday (17 July 2019, 8am–10am), the number of people in this two-hour flow record has reached 286. And in the selected summer weekend (20 July 2019, 8am–10 am), the number of people in this two-hour flow record has reached 359. Another characteristic of research data is related to the type of people's outdoor activities—optional activities and social activities occur with consistent frequency in the 3rd community centre. Shopping (*See* Fig. 5.9), standing and chatting (*See* Fig. 5.10), as well as sitting and resting have been seen in each recorded period, however, people's average staying time all lasts less than 30 mins, which means they stay in this place not for a long time.

According to the evaluation of people's activities and spatial quality, commuting is less affected by the environment, however, the good-quality slow-traffic system could induce purposeful necessary activities and could drive social activities (Sun et al., 2017, p. 3). By comparing these two poor-spatial public spaces, observed phenomenon seems to

**Fig. 5.8** People walking and waiting in Zhongxin Avenue (*Source* Figures are taken by the researcher. 15 July 2019)

**Fig. 5.9** Shopping in the supermarket of 3rd community centre

imply an 'invisible voice' to the researcher—people do not want to spend more time in the poor-quality public space as long as they complete their necessary daily activities.

As discussed about elements that constitute responsive public spaces of the SSTEC, the element of variety mostly relates to the integration of land use in a manner that promotes ease of access, that is, being located along the major transportation networks (vehicular and pedestrian) in such a way that they provide opportunities that are easily accessible to a range of users. The provision of a range of uses contributes to the complexity, and the measure of complexity would relate to the extent of uses that are available to people. The evaluation of robustness is to ensure the place can support a wide range of uses. Examples of this element referring to the SSTEC case are more related to the 3rd community centre which can be used as daily shopping places and a meeting place for people, as well as children's park where the adult can relax and enjoy the shade while looking after their children, and the playlot for children is in the same place.

Through the comparative analysis of people's using time and using frequency of both good-quality and poor-quality public spaces within the eco-city, conclusion of this section has been summarized as follows. The representation of *people–place interactive relationship* concept in the real-world planning project is interpreted as the behavioural connection between people and place—people indirectly express their satisfaction or

**Fig. 5.10** People standing and chatting in 3rd community centre (*Source* Figures are taken by the researcher)

dissatisfaction with spatial quality by their outdoor activities. In good-quality public space, people's 'eyes on the public space' has provided informal surveillance to the surrounding environment. Also, they are willing to spend more time for staying, socializing, actively interacting with others, and different age groups can enjoy themselves. While in poor-quality public space, people express their dissatisfaction by a kind of seemingly fleeing behaviour.

## 5.6 Safety Issue: Core Element of the Behavioural Connection

*Environmental Psychology* mainly discusses the relationship between urban environment and residents' psychology, as being revealed in Chapter 2, one of the obvious goal and principle of the *Environmental Psychology* is about considering the person and the setting to be a holistic entity. Moreover, it examines the way that people conceptualize natural environment, and how people imagine more comfortable and pleasant places. Combined with the *New-Type Conceptual Model of Human-Centred Design* shown in Chapter 3, human-focused design intends to satisfy people's higher level of needs, namely the psychological aspect such as cognitive and aesthetic aspects. This section aims to explore the consideration of people's sense of activity experience, namely the core element that influences people's psychological feeling in public spaces. It is related to research methods of questionnaire and street-intercept interviews.

### 5.6.1 *Safety of Poor Spatial Environment: Zhongxin Avenue*

In terms of the initial analysis of questionnaire, the majority of answers to *Question 14, Besides good and bad, can you please write three adjectives to describe the overall spatial environment of this research site,* tend to be the negative adjectives, namely unsafe, uncomfortable and the desolate. Referring to Zhongxin Avenue research area, among 78 questionnaires, 75 copies have been pointed out to people's feeling towards the spatial quality related to the safety adjectives, such as *unsafe*, *uncomfortable* and *desolate*. Such statistics indicate that people regard safety issue as the primary consideration of their outdoor activity experience, influencing people's psychological feeling in the public space. This has been also verified in the data analysis of two selected street-intercept interviews conducted in Zhongxin Avenue.

The first interview transcript clearly indicates the absence of protective measures, such as isolation railings or ornamental plants, to prevent pedestrians from coming into contact with vehicles, makes her feel unsafe. This lack of protection is the primary factor contributing to her discomfort and inability to enjoy outdoor activities along Zhongxin Avenue, which forms part of her jogging route. By coding the second transcript of interview, it can be seen that apart from lack of protective approaches (*See* Fig. 5.11), traffic issue about the narrow sidewalk (*See* Fig. 5.12) has been seen as another reason causing people to feel uncomfortable and unsafe. Specifically, narrow, to some extent, means not walkable and not in human scale, people may feel negative psychological feelings (e.g. dangerous and uncomfortable listed in questionnaires). Known from the interview, the elderly people usually take their grandchildren to school and the adults usually take bus to workplace. With too many people crowded in such narrow street without any traffic protection, most of them may feel dangerous as cars could come speeding through at any moment. As the interviewee describes, traffic environment is unfriendly to the elderly and children. In other words, such road design does not follow people-oriented principle, that is, it does not put people's fundamental safety needs in the first place.

In summary, the case of jogging has indicated the relationship between people's psychological feeling and their behaviours in public space—if place cannot make people feel safe, people will not enjoy activities in such place as uncomfortable feeling can stop them experiencing enjoyment. Moreover, the case of narrow sidewalks in Zhongxin Avenue has also indicated that the sense of safety, an important design principle of people-oriented planning, is always the most concerned aspect for people doing activities in public space. That is, safety has been regarded as the fundamental need for people's public lives.

**Fig. 5.11** Lack of protective approaches in Zhongxin Avenue

**Fig. 5.12** Narrow sidewalk in Zhongxin Avenue (*Source* Figures are taken by the researcher)

### 5.6.2 *Safety of Good Spatial Environment: Children's Park*

In terms of the initial analysis of questionnaire, the majority of answers to *Question 14, Besides good and bad, can you please write three adjectives to describe the overall spatial environment of this research site,* tend to be all positive adjectives, such as *safe*, *happy*, *comfort*. For the children's park, among 85 questionnaires, 82 copies have pointed out people's feeling towards the spatial quality related to the good safety conditions, such as the adult has reasonable viewing distance to keep eyes on their children, as well as people's activities happen at all hours of the day. In addition, 78 copies have used 'happy' adjective to describe their pleasant feeling when staying in this place—children can enjoy playing a variety of infrastructures and can make new friends, meanwhile, adults can talk with each other and can share their daily lives when looking after their children. Such phenomenon has also been used to describe good views of this place as adults can easily keep an eye on their children playing nearby due to the reasonable viewing distance.

From the four selected street-intercept interviews, it is obvious that people's psychological feeling towards children's park are all positive, even though they are in different age groups. Most importantly, all these four interviewees used *I feel safe* before they express their satisfaction towards this public space. Specifically, information obtained in Transcript 1 shows that, the interviewee's grandson is *eager to play with his friends here, despite the high temperature*. It seems to imply this children's park has been regarded as a 'magnetic field' and has great attraction to children. Also, the interviewee in Transcript 2 has used the sentence *it is so interesting and I can enjoy myself here* to express her feeling of affection towards the spatial quality of this public space. In Transcript 3, the interviewee has briefly introduced her and her family members' routines by using this public space and has expressed her happy feeling for the retired life. Based on her description, this warm family scene has indirectly told the researcher about their attachment to this public space. In addition, in Transcript 4, the interviewee has indicated the popularity of this public space—there are quite a lot of people here during the peak hours (e.g. 9–10am and 6–8pm).

① *My grandson especially likes this place. It is summer vacation, he is eager to play with his friends here every morning, despite the high temperature. I often do shopping in the nearby supermarket (the 3rd community centre) and take him to play here about one hour, then go home. (Resident, female, about 65 years old) (Interview time: 10.20am, 16 July 2019).*

② *It is so interesting and I can enjoy myself here. My classmates and I often come here to play games after class. I can also make new friends here after playing with them. However, I think more sports facilities should be provided here, as our boys like playing basketball and riding skateboard. (Resident, girl, 10 years old) (Interview time: 1.30pm, 14 July 2019).*

③ *My husband and I are used to come here after dinner. He plays with our granddaughter, and I participate the square dancing with my neighbours (middle-aged or elderly women). We go home together before 9am, and the retired life is so happy. (Resident, female, about 60 years old) (Interview time: 6.15pm, 16 July 2019).*

④ *Only the third community is equipped with the children's park, even some parents who live in the first and second communities often take their children to play here. There are quite a lot of people here during the peak hours (e.g. 9–10am and 6–8pm). But I am wondering why there are no children's park in the first and second communities? (Resident, female, about 35 years old) (Interview time: 10.40am, 16 July 2019).*

### 5.6.3 *Safety Issue: The Precondition to Make People Psychologically Feel Comfortable*

As shown in the evaluation criteria used in the pilot and the 12 quality criteria, proposed by Jan Gehl, have been regarded as the guideline for creating and designing public spaces. Safety issues are divided into three categories, in the name of protection, including protection against traffic and accidents (feeling safe), protection against crime and violence (feeling secure) and protection against unpleasant sensory experience. Thus, emphasis about the initial data analysis in terms of the safety issue have been placed on the above three aspects.

Referring to elements that constitute responsive public spaces in Chapter 5, the purpose of permeability and legibility is to expand the routing and to improve people's ability to orientate themselves within the place. In this research, permeability and legibility, as two of responsive indicators, especially refer to a transit route without obstacles or a route that allows people to saunter with sense of safety through their surroundings.

Theoretically, already mentioned in Chapter 2, Jane Jacobs has discussed the significance of safety in the public space in her book *The Death and Life of Great American Cities*. She introduces the crime-preventive effect of public life and people's care for common space. Her expressions, like *street watchers* and *eyes on the public space* have become integral to urban planning terminology. As Gehl (2010) argues, 'being able to walk safely in city space is a prerequisite for creating inviting well-functioning places for people. Experienced as well as perceived safety is crucial for public life in the city' (p. 97). The safety discussion has a general and a more detailed dimension. The general focus is maintaining and supporting the vision of comfort environment that people with different age groups can do daily activities. Within this general framework, safety can also be promoted through careful considerations for detailed designing solutions, such as natural surveillance and pedestrian protection, in order to realize behavioural connection between people and place in the (public) space.

Practically, people's perceived safety, their behaviours and the built environment are interconnected. By comparing research data of Zhongxin Avenue and children's park, conclusion can be summarized as follows. Safety issue has been seen as the precondition to make people psychological comfort and delight. That is, people regard the safety issue as the primary consideration of their sense of activity experience, significantly influencing their psychological feeling in public spaces—unsafe environment makes people hold negative attitudes, such as uncomfortableness, towards spatial quality of this public space (e.g. Zhongxin Avenue), vice versa (e.g. children's park). That is, as *Maslow's Hierarchy of Needs* reveals, the need for basic safety should be fulfilled when pursuing higher-level needs.

## 5.7 Social Infrastructure: Means of Realization for the Interactive Connection Between People and Place

Social infrastructures are of great importance during the eco-city development and operation stages. Functioned as the theoretical foundation, the guidance of Garden City could be seen as the demonstration effect of eco-city development. Specifically, in the *Nine Garden City Principles* listed by Town and Country Planning Association (*See* Table 2.5), the eighth principle is about the strong cultural, recreational and shopping facilities in walkable, vibrant, sociable neighbourhoods. Caprotti and Gong (2017) made a case for 'humanizing' newly built eco-cities by emphasizing on urban social sustainability through the lens of experiences of new residents. They argued 'there is a need to focus on the ways in which socially sustainable urban environments are constructed, in new urban spaces, through relational networks comprised by interactions between residents, buildings, facilities and specific (e.g. domestic) spaces' (p. 45). This section aims to explore by what means one can achieve this interactive connection between people and place, namely how does the design concept of interaction embody in real-world scene of public space from the perspective of environmental elements. It is related to research methods of questionnaire and street-intercept interviews.

### 5.7.1 *Social Infrastructure of Good Spatial Environment: Eco-Valley*

In terms of the initial analysis of questionnaire, the majority of answers to *Question 14, Besides good and bad, can you please write three adjectives to describe the overall spatial environment of this research site*, tend to be all positive adjectives, such as *safe*, *convenient*, *comfort*, *happy* and *good views*. Among 132 questionnaires, 109 copies have used *convenient* and 102 copies have used *comfort* adjective to describe pleasant social infrastructures—well-equipped jogging track (*See* Fig. 5.13) and good primary seating options (*See* Fig. 5.14). Specifically, the elderly can enjoy the sunshine and can chat with other neighbours in the resting area with good primary seating options, especially for those retired people who love outdoor activities with others (*See* Fig. 5.15). Young people can relax themselves after a long-day work by using the well-equipped jogging track.

**Fig. 5.13** Good jogging track in the eco-valley (*Source* Figures are taken by the researcher)

**Fig. 5.14** Good primary seating options in the eco-valley (*Source* Figures are taken by the researcher)

### 5.7.2 *Social Infrastructure of Poor Spatial Environment: The 3rd Community Centre*

In terms of the initial analysis of questionnaire, the majority of answers to *Question 14, Besides good and bad, can you please write three adjectives to describe the overall spatial environment of this research site,* tend to be the negative adjectives, such as *uncomfortable* and i*nconvenient*. Among 138 questionnaire, 112 copies have pointed out people's feeling towards the spatial quality is related to the sense of comfort caused by the shortage of seating. It is worth mentioning that among these 112 copies, 73 copies have stated people feel hot in summer and cold in winter due to the lack of sheltered resting space. In addition, among the total 138 questionnaire,

**Fig. 5.15** The elderly enjoy the sunshine and chat with others (*Source* Figures are taken by the researcher)

85 copies have used *inconvenient* adjective to describe issues of social infrastructures within the 3rd community centre. For example, issues about the lack of entertainment and catering service for young people as well as the lack of barrier-free access (*See* Fig. 5.16) at entrance for the elderly people and women with children, making different age levels of people feel inconvenient when staying in this public space. Research data conducted in the 3rd community centre has indicated that people's psychological feeling about the spatial quality of public space is directly reflected by the service conditions of basic social infrastructures. Such

conclusion has been also verified in the data analysis of street-intercept interviews. Below is the detailed description of these selected interviews.

① *The opening hour of here (indoor market) is from 7am to 8pm, my wife and I come to arrange the market stall at 6.30am every day. However, the air conditioning supply time is from 8am to 7pm. For us (stallholders), there is always a period of uncomfortable feeling, namely hot in such summer time and cold in winter time. (Stallholder, not resident, male, 35 years old) (Interview time: 9am, 17 July 2019).*

② *I live in the nearby community, walking to here (indoor market) within 15 mins. I come here twice or three times a week, and usually come between 9am to 9.30am. It is so convenient for our elderly people. … I remember that it was rebuilt 3 years ago. Nowadays, the tile floor has become tidier and space between stalls has become much wider, I feel much safer to use my trolley to buy food here. In addition, the most*

**Fig. 5.16** The elderly lifts baby carriage when climbing stairs (*Source* Figure is taken by the researcher)

*comfortable thing in this indoor market is to have air conditioners—not hot in summer and not cold in winter. (Resident, female, about 70 years old) (Interview time: 9.20am, 17 July 2019).*

③ *I think some facilities here (the entrance of community centre) are actually not convenient. Our elderly people like to chat with neighbours after grocery shopping, however, there is no resting area here. We will prefer to spend our more time here if there is less flowers and trees, but more resting seats. Otherwise, we often directly go back home by bus after buying food. (Resident, male, about 70 years old) (Interview time: 10am, 17 July 2019).*

④ *I have been working this bakery for more than one year. I am not live in this eco-city, so I come here by bus every day. As a work place, it (the community centre) should be safe and convenient for transport. It indeed lacks the resting area. I quite like to stay outside for relaxing during the lunch break when sunny days. (Employee, non-resident, male, 35 years old) (Interview time: 9.15am, 20 July 2019).*

⑤ *I often come here to buy food with my child (baby boy, about 1 year old), and sometimes his grandparents are carrying him with baby carriage. Our family feels that here (the entrance of community centre) lacks barrier-free access. Otherwise, all adults carrying their children need to move our baby carriages up and down. I do not like here as it is so inconvenient for us, especially when buying more food. (Resident, female, about 30 years old) (Interview time: 10.40am, 20 July 2019).*

⑥ *We are both undergraduates and come here for job hunting in the summer vacation, and have been here for more than 10 days. We think the whole eco-city lacks the entertainment and catering service for our young people. As we do part-time job near this community centre, we have to come here eating KFC for lunch or call for a takeout in our office. As young people, we prefer a variety of catering services, such as bubble tea shop, café and BBQ restaurant. … These days, we have to sit here (steps of entrance) or stay in KFC for resting a while during our lunch time. As the temperature is so high in the summer time, we indeed feel really uncomfortable. We hope to have more places in the eco-city for the young people to rest, like Starbucks. (Non-residents, two females, each is about 20 years old) (Interview time: 1 pm, 20 July 2019).*

Take the indoor market in the 3rd community centre as an example. Data conclusion tends to be contradictory. Information obtained from Transcript 2 shows that air-conditioning supply time in the indoor market is usually one-hour later than the opening time and one-hour earlier than the closing time. Such supply time has little impact on residents, as data shown in Transcript 2, the interviewee usually come to the market approximately two hours later than the opening time, during which the air-conditioning has been already supplied for an hour. Thus, she thinks this market is convenient for elderly people as this place makes her feel comfortable due to the suitable indoor temperature for people's staying. On the contrary, information obtained from Transcript 1 shows that, being a stallholder, the interviewee usually comes to the market with his wife for preparation half an hour earlier than the opening time, during which the air-conditioning has not been supplied. In addition, as the air-conditioning stops supplying one-hour earlier than the closing time, it always makes those stallholders feel hot in summer and cold in winter (uncomfortable). The uniform air-conditioning supply time has brought different sense of comfort to different people, making people hold different attitudes towards the place.

On the other hand, referring to the sense of comfort and convenience in the outdoor environment, people hold quite same attitudes. Take the entrance of community centre as an example. Based on *Transcript 3* and *Transcript 4*, data conclusion tends to be consistent. Both resident (interviewee from *Transcript 3*) and non-resident (interviewee from *Transcript 4*) have pointed out the reason which prevents them from staying more time in the outdoor space owns to the shortage of seating options for people to chat and to have a rest. People have to sit on the stone steps at the edge of landscape boundary or sit on their own chairs that they bring from home. Phenomenon of inconvenient social infrastructures also reflects in *Transcript 5*. As the interviewee demonstrates, because of lacking barrier-free access, adults who are carrying their children have to move their baby carriages up and down. Also, the interviewee has used the sentence *I do not like here as it is so inconvenient* to express her dislike feeling about the spatial quality of this public space. In addition, in *Transcript 6*, two non-residents describe they feel uncomfortable as during high-temperature summer time, they cannot find a pleasant place for young people to have a rest, such as bubble tea shop, café or BBQ restaurant (Fig. 5.17).

**Fig. 5.17** Shortage of seating options at the entrance (*Source* Figure is taken by the researcher)

Referring to elements that constitute responsive public spaces in Chapter 5, the purpose of variety in public space is to create a physical setting for the well-functioned space (Talen, 2002, p. 167). Murrain (1993) regarded variety as 'a finely grained mix of primary land uses, being supported by public services' (p. 171). In this research, variety as one of responsive indicators especially refers to the variety of social infrastructures available to the variety of users. Research data conducted in the 3rd community centre shows the direct cause of people's positive or negative feeling is precisely due to the condition of social infrastructures. People feel comfortable or convenient if the indoor environment has well air-supplying condition, while people feel uncomfortable or inconvenient if the outdoor environment has inadequate seating options and barrier-free access.

The connection between human activity and the physical space, often known as social infrastructure, is the third component in the interactive set of elements that make up the content of urban design. Social infrastructure, or public service facility, is 'an important element to show design details and the degree of humanization, also is the basic condition for building the good atmosphere of public space' (Sun et al., 2017, p. 5). It not only shapes characteristics of public spaces, but also provides guarantees for the occurrence and continuation of people's outdoor activities. Based on the above data analysis, social infrastructures such as resting areas and barrier-free facilities have been seen as another primary consideration of their sense of activity experience. The *Report of the Implementation of Indicator System of the Sino-Singapore Tianjin Eco-City (2020)* points out 'the indicator stipulates that by 2023, the coverage of barrier-free facilities in public space shall reach 100%, the index reflects the SSTEC's people-oriented planning concept'. Nevertheless, residents still have some dissatisfactions about the basic living facilities in the eco-city.

### 5.7.3 *Flow and Fixity to Understand How People Experience Places*

Diverse forms of mobility reconfigure how people experience places. Since the early 2000s, the systematic exploration of urban mobility has become centre of an emerging interdisciplinary research field, variously describing as *the mobilities turn* or *the new mobilities paradigm* in social science discipline (Cresswell, 2006; Urry, 2000). However, as Urry (2000) criticized, for spatial ordering, 'social science has failed to fully recognize how the spatiality of social life presupposes, frequently involve conflict over, both the actual and the imagined movement of people from place to place, event to event' (p. 4). In recent years, literatures have given increased attention to the role of people–place bonds in various realms of mobility, including residential mobility, tourism-work-amenity-seeking migration (Buckle, 2017; Easthope, 2009; Haynes, 2008; Williams and McIntyre, 2012).

Masso (2019) provided a rational and conceptual framework (*See* Fig. 5.18) for research on place attachment that attends to focus on more fluid experiences of place endemic to a mobile world. Table 5.4 shows modes of interrelation and configurations of fixity-flow shaping place attachment. As a response to the mobilities turn, a new framework has been built around the conceptual spectrum from fixity to flow that re-theorizes and reorganizes different modes of interrelation between fixed and fluid aspects of place attachment dynamics—'between the more spatially static and temporally stable aspects of place attachment as a topologically centred experience (i.e. fixity), and the more spatially mobile and temporally changing aspects of place attachment as topologically centreless experience (i.e. flow)' (Masso, 2019, p. 128). Between the extreme poles of fixity and flow, the framework displays how the interrelation of the two can shed light on mobility-driven place attachments. Such framework represents 'a conceptual spectrum of modes and not a sequential one' (*ibid*), and these modes should be understood as different manifestations of fixity-flow with permeable boundaries.

Compensation between fixity and flow posits 'mobilities might enable a form of place attachment that makes up for unsatisfactory fixed place-bonds' (Masso, 2019, p. 129). Studies have shown how people made up for aspects of place attachment lacking in the home environment either by engaging in leisure mobility (Fuhrer et al., 1993) or bonding with non-residential places (Manzo, 2005). Fixity-flow framework proposes the theoretical consideration of place attachment which is more 'fluid and relational' (Masso, 2019, p. 130) compared to Massey's (1994) notion of the global sense of place which has been well-acknowledged by human geography (Cresswell, 2015; Williams, 2014). Implications of such approach to place attachment supports a view of place being identified as assemblages (Cresswell, 2015), which spread feelings of anchoring and rootedness across multiple and dynamic locales (Williams & McIntyre, 2001). As Masso (2019) stated, place attachment has been regarded

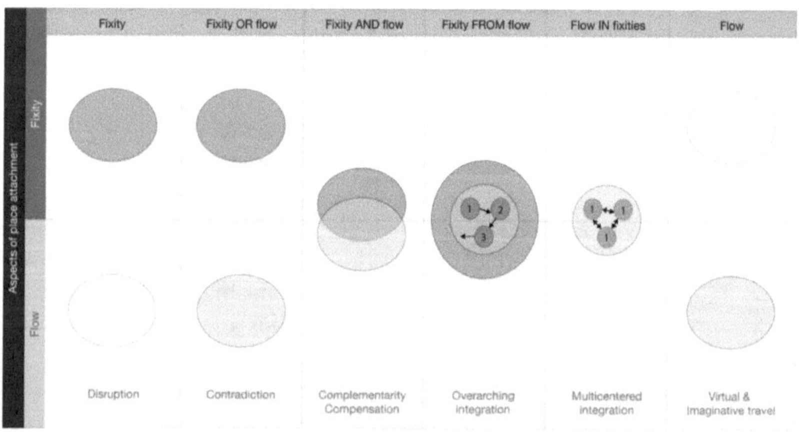


**Fig. 5.18** Conceptual framework with main modes of interrelation between the fixed and the fluid aspects shaping place attachments (*Source* Figure is from Masso, 2019, p. 128.)

**Table 5.4** Interrelation and configurations of fixity-flow attachment

| *Mode of interrelation* | *Configuration* | *Description* | *Example studies* |
|---|---|---|---|
| Fixity | Disruption | Mobility and place attachment represent two opposing and mutually exclusive concepts—one must choose between staying or leaving | Brown and Perkins (1992), Fried (1963), Fullilove (2001), Manzo (2014a) |
| Fixity or flow | Contradiction | Places offer distinct yet complementary experiences: they allow a balance between the desire and capacity to travel while maintaining a strong attachment to one's home, or the active formation of new place connections when relocating | Gustafson (2001b) |
| Fixity and flow | Complementarity Compensation | Mobility can fill the gaps that fixed place bonds might leave unfulfilled. It may not provide local knowledge or sense of self-affirmation associated with a lifestyle deeply rooted in one place | Gustafson (2009b), Lewicka (2013), Bailey et al. (2016), Stedman (2003), McIntyre et al. (2006), Williams and van Patten (2006), Fuhrer et al. (1993), Buchecker (2009) |
| Fixity from flow | Overarching integration | Movement between locations may evoke an overall sense of stability, especially if: (a) there is a feeling of continuity across similar geographic areas, or (b) there is alignment with one's sense of self, including self-continuity, self-efficacy, distinctiveness, security, or aesthetic fulfillment | Williams and Kaltenborn (1999), Williams and McIntyre (2001) |

(continued)

**Table 5.4** (continued)

| *Mode of interrelation* | *Configuration* | *Description* | *Example studies* |
|---|---|---|---|
| Flow in fixities | Multi-centred integration | Temporary or shifting connections between places form a network of meaningful nodes or rhizomatic links that resonate with one another, where the individual feels attached, identified | Szerszyki and Urry (2006) |
| Flow | Virtual travel<br>Imaginative travel | This transcendence of physical and often social distances is facilitated by information<br>Through the media, individuals are virtually transported to other places, experiencing images of locations and people from afar | Champ et al. (2013), Szerszynski and Urry (2006) |

*Source* Information is from Masso, (2019, p. 129). Summarized by the author

as 'an emergent property of a complex system' (p. 131), and within this system, people adapt and evolve new repertoires of behaviour, 'effectively making and remaking the system through action' (Williams, 2018, p. 289).

Designing public spaces that accommodate both flows (movement and activities) and fixity (comfort and sense of place) requires a thoughtful approach that balances these two seemingly opposing aspects to support place attachment. For the case of Zhongxin Avenue, the space is sometimes used mostly by people walking between bus stops or similar uses. It is more of a space of mobility and flow, than a space which welcomes people stopping, chatting and staying in space, which makes the Zhongxin Avenue become different compared to other selected research locations. The reason for this selection aims to discuss how to design public spaces that accept flows and fixity at the same time, in order to make poor-quality public spaces within the eco-city become more humanistic and desired for users' experience. On the one hand, to provide enough seating and rest areas at some locations by encouraging people to interact with the surrounding environment, and the bus station is a good location choice.

On the other hand, as discussed in Chapter 4, the development of SSTEC is still an eco-city, but now more of a smart and/or digital city in focus. To adopt smart technologies for traffic management and energy-efficient lighting, these systems can adapt to real-time flow dynamics and to make people's travel more convenient while maintaining the city's ecological balance.

## 5.8 Data Collection and Discussion During Supplementary Fieldtrip

### *5.8.1 How Safety and Social Infrastructure Are Related to the Evaluation of Public Space*

This research insists creating people-centred spatial environment requires an intersecting holistic approach, not only limits to the eco-city development but also should expand to the whole planning concept. The relationship between safety and social infrastructures are complementary to each other, namely the complementary sequential relationship. Specially, when the public space has been designed with safety in mind, it creates the environment where people feel safe, enjoyable and comfortable in using such place. On the premise of meeting people's needs towards safety, the spatial environment that makes people feel satisfied will naturally gather the popularity and also engage people to participate in more outdoor activities. At this time point, it is necessary for social infrastructures to take actions to meet people's changing needs in different periods by relying on the reform of updated digital technical means. Further research data has been collected in June 2023 during the supplementary fieldtrip to the SSTEC.

In terms of the consideration of safety, the children's park is a case in point. During the summer observation in July 2019, children raced across the playground, a group of parents as well as caregivers sat on nearby benches, engrossed in conversation while keeping a watchful eye on their little ones. While in the summer observation in June 2023, the same scene still plays out in the children's park (*See* Fig. 5.19). One in-depth interview conducted on 16 April 2023 with the general leader of L&A Design, who is responsible for the SSTEC community park project design, demonstrated how children's park penetrates the people-centred planning and design concept (esp. the children-friendly planning concept) into safe and high-quality planning version. As the transcript of interview shows, the reason why children's park has been still popular with eco-city's residents is due to the careful consideration of design details for children's safety.

> The construction of people-oriented and sustainable open space largely depends on whether it is symbiotic with urban ecology and surrounding community life. For children's park, the sustainability of the SSTEC project, includes not only the use of ecological technology, but also the maintenance and operation of the park. Building safe and resilience strategies include not only urban ecological resilience but also multi-dimensional adaptability to community life, which creates a concentric circle of humanistic planning in the context of ecology, green ecology and contemporary people's needs. Based on children's walking habits and taking the community as a unit, the children's park has integrated children's planning scenes into major public spaces, so that children can play happily and enjoy nature anytime and anywhere. ... I will show you some examples of design details - the ground of the children's park is all made of rubber mats to prevent children from failing and bumping, and social facilities are avoided sharp angles. The supporting service facilities are designed according to the physiological scale of the child, and the children's park has provided shaded porch pavilions as well as tables and chairs for accompanying parents. (Mrs. Li, the general leader of L&A Design, interviewed on 16 April 2023)

Variety as one of responsive indicators especially refers to the variety of social infrastructures available to the variety of users. The direct cause of people's feeling due to the condition of social infrastructures—people feel uncomfortable or inconvenient if the outdoor environment has inadequate seating options and barrier-free access. However, during the supplementary fieldwork conducted at the end of June 2023, the 3rd community centre still lacks seating facilities. And based upon the interview with the young man, he expressed his dissatisfaction with the lack of seating facilities by comparing ones in the 4th community centre (Figs. 5.20, 5.21, 5.22).

> To allow people to stop and rest, the smart seat can also be connected to a 5G network, and is equipped with USB wired fast charging, wireless charging and other functions. When I was tired from walking or exercising, I can also charge my phone while sitting. Such function is very friendly to people. (Residents, male, 25 years old)

**Fig. 5.19** Playing and chatting scene in June 2023 (*Source* Images are taken by the researcher)

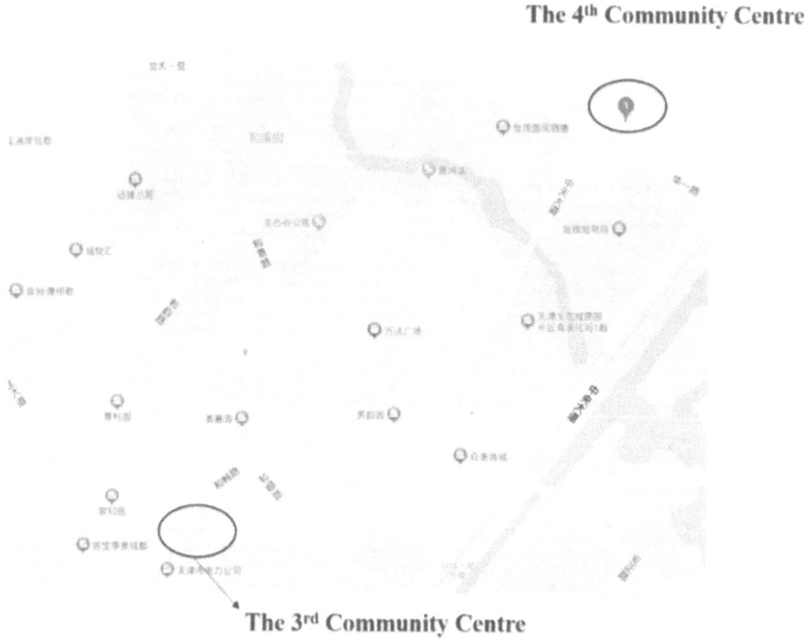


**Fig. 5.20** Locations of the 3rd and the 4th community centre

> In addition to the convenience of shopping, the most of favourite thing for my daughter and for me in the 4th community centre is the brand-new seating facilities, which make us be willing to stay more time, regardless of the high temperature in summer. (Residents, female,36 years old)

### 5.8.2 *How People–Place Relationship at the Micro-Scale Is Linked to the Broader Humanistic Planning Aspect at the Macro Scale*

This section aims to dive deeper into the relationship between people–place interaction at the micro-scale and how they inform humanistic planning at the macro-scale for eco-city planning. It focuses on user-centred design, sense of place, social inclusion, environmental sustainability and people's quality of life. Detailed explanation has been shown as follows.

Firstly, humanistic planning recognizes the importance of designing spaces that lay emphasis on users' needs and behaviours. By studying how

**Fig. 5.21** Seating facilities in the 3rd community centre

individuals interact with specific places at the micro-scale, planning practitioners can gather first-hand data on usage patterns and preferences, and then they apply such information to macro-scale planning decisions, such as determining the placement of amenities, designing social infrastructures or shaping the layout of neighbourhoods. This research holds a view that humanistic planning ensures the built environment is functional, accessible and comfortable to people by prioritizing using needs of individuals. Secondly, people develop strong psychological connections (e.g. safety, enjoyment) to places they live, and these are influenced by some criteria of responsive environment, such as permeability and legibility, variety and robustness. The humanistic planning approach recognizes the importance of enhancing spatial qualities at the micro-scale to create a distinct sense of place at the macro-scale, and such people-centred planning approach helps to create vibrant places that contribute to the overall attractiveness and liveability of places. Thirdly, the humanistic planning approach recognizes the importance of fostering social interaction and diversity within the built environment, thus, the micro-scale interaction between people and place has impact on community well-being. By observing

**Fig. 5.22** Seating facilities in the 4th community centre (*Source* Images are taken by the researcher)

how individuals gather, socialize and engage with their surroundings at the micro-scale, planning practitioners can formulate macro-scale eco city planning strategies to promote the concept of humanistic planning under the context of ecology, by creating public spaces that accommodate a variety of activities, or ensuring the built environment is accessible to people regardless of their ages and genders. Fourthly, the micro-scale relationship between people and place also plays a crucial role in addressing environmental challenges at the macro-scale, especially for the eco-city planning. By understanding how individuals use and navigate their surroundings, planning practitioners can formulate macro-scale eco-city planning strategies to encourage people's sustainable behaviours such as identifying opportunities to improve public transit systems to make green public transport become people's first choice, or creating bike-friendly infrastructures to promote pedestrian-friendly environment. In short, by prioritizing the well-being of individuals, the humanistic planning under the context of ecology aims to create the built environment that is not only functional and sustainable, but also enriching and enjoyable to people.

## 5.9 Conclusion

The ultimate goal of the SSTEC is to 'create a city that coexists in harmony with the natural society' (Yang & Dong, 2008, pp. 246–247) which should not only meet requirements of the regional ecosystem and build the space that combines its own ecological characteristics, but also provide good physical environment for residents. Fit, one of five performance dimensions to measure good physical environment, refers to the match between people's behaviours. It is 'how well the spatial and temporal pattern of a settlement matches the customary behaviour of its inhabitants' (Lynch, 1984, p. 151). This chapter illustrates the relationship between people and place from the perspective of residents of the eco-city. As being identified, people's behaviours are the connecting format between people and place, people's behaviours will be used as a significant component to further discuss how can people–place relationship be introduced or supported in the Chinese eco-city planning practices. Research data, collected by means of pilot study and participant observation at the main fieldwork stage, have identified how does the *people–place interactive relationship* embody the core connotation of people-oriented concept in the eco-city development. This chapter has also revealed safety issue as the precondition to make people psychological feel comfortable, and social infrastructure ties people and place together—these are two forms of the behavioural connection between people and place, and also are people's fundamental needs when staying in the public space within the eco-city.

A key dimension of a good city is its collective public character. Klinenberg (2018) argued that a whole range of physical and institutional infrastructure are of importance for the development and maintenance of social connections. These places, such as parks, sports facilities and community centres, indeed matter as they are sites where people can meet and 'mix with others with whom they share their neighbourhoods' (Latham & Layton, 2019, p. 2). For Klinenberg, social infrastructures are not only necessary for nurturing public life, but also for addressing some concerns of contemporary urban life, such as countering social isolation and creating places for all—regardless of age, race, gender or income. Facilitating sociality is an essential component of how social infrastructures manage to provide their primary functions—social infrastructures may exist to amplify connections within different groups, also, they help to orientate people towards interacting (Blommaert, 2014). As Latham and Layton (2019) identified, social infrastructure refers to 'the network of spaces and facilities that create affordances for social connection' (p. 3). For the case of 4th community centre, the social infrastructure is understood as an integral part of urban fabric, namely the background technological networks and systems that support urban life. At the conceptual level, the term infrastructure is useful as it is the 'background structures and systems that allow social life to happen' (Latham & Layton, 2019, p. 3). With infrastructure, the central dynamic is around the facilitation of people's activities. Social infrastructures involve considering different kinds of facilities necessary for eco-city to function as social spaces as facilities serve distinct functions. For the study sites of SSTEC, the children's park and the eco-valley provide space for people of all ages to exercise, especially for children and the elderly. The 3rd community centre and the 4th community centre mainly provide commercial services to residents. These public spaces are accessible localities where people socialize and make connections with others, and they all have significant functions that residents make use of while living.

As already discussed in Chapter 4, by taking the consideration of residents' experience as an example, the revised indicator system (2.0 version) has increased the proportion of residential areas within a five-minute walk from parks and green space. And it also has increased the proportion of indicators that can directly reflect the improvement of people's sense of well-being, such as indicators of residents' community activities, residents' life satisfaction and residents' health literacy level. Nevertheless, residents still have some dissatisfactions about different needs physically and psychological when living in the eco-city, which has been seen as the conceptual-practical gap of the people-oriented concept. From this perspective, the next chapter will firstly explore causes of such dissatisfaction by contrasting the residents' feelings and designing intention of planners, in order to explore whether the people–place interactive relationship being regarded as the representation of people-oriented planning goal in the eco-city project can reduce such performance difference.

## References

Aspinall, P., Mavros, P., Coyne, R., & Roe, J. (2015). The Urban Brain: Analysing Outdoor Physical Activity with Mobile EEG. *British Journal of Sports Medicine, 49*(4), 272–276.

Bailey, E., Devine-Wright, P., & Batel, S. (2016). Using a Narrative Approach to Understand Place Attachments and Responses to Power Line Proposals: The Importance of Lifeplace Trajectories. *Journal of Environmental Psychology, 48*, 200–211.

Blommaert, J. (2014). Infrastructures of Superdiversity: Conviviality and Language in an Antwerp Neighbourhood. *European Journal of Cultural Studies, 17*(4), 431–451.

Brown, B. B., & Perkins, D. D. (1992). Disruptions in Place Attachment. Cited In: Altman, I. and Low, S.M. (1992). *Place Attachment: Human Behaviour and Environment* (pp. 279–304). 1st ed. Springer.

Buchecker, M. (2009). Withdrawal from the Local Public Place: Understanding the Process of Spatial Alienation. *Landscape Research, 34*(3), 279–297.

Buckle, C. (2017). Residential Mobility and Moving Home. *Geography Compass, 11*(5), Article e12314.

Caprotti, F., & Gong, Z. Y. (2017). Social Sustainability and Residents' Experiences in a New Chinese Eco-City. *Habitat International, 61*(2017), 45–54.

Catharine, W. T. (2011). Linking Landscape and Health: The Recurring Theme. *Landscape and Urban Planning, 99*(3/4), 187–195.

Champ, J. G., Williams, D. R., & Lundy, C. M. (2013). An On-line Narrative of Colorado Wilderness: Self-in- Cybernetic Space. *Environmental Communication, 7*, 131–145.

Cooper M, C., & Francis, C. (1998). *People Places: Design Guidelines for Urban Open Space*. 1st ed. John Wiley & Sons.
Cresswell, T. (2006). *On the Move: Mobilities in the Modern Western World* (1st ed.). Routledge.
Cresswell, T. (2015). *Place: An Introduction*. 2nd ed. Wiley Blackwell.
Cullen, G. (1961). *The Concise Townscape* (1st ed.). Routledge.
Easthope, H. (2009). Fixed Identities in a Mobile World? The Relationship Between Mobility, Place and Identity. *Identities: Global Studies in Culture and Power, 16*(1), 61–82.
Fried, M. (1963). Urban Problems. *American Journal of Orthopsychiatry, 33*(2), 342–345.
Fuhrer, U., Kaiser, F. G., & Hartig, T. (1993). Place Attachment and Mobility During Leisure Time. *Journal of Environmental Psychology, 13*(4), 309–321.
Fullilove, M. T. (2001). Root Shock: The consequences of African American dispossession. *Journal of Urban Health. 78*(1), 72–80.
Gehl, J. (1987). *Life Between Buildings: Using Public Space* (2nd ed.). Van Nostrand Reinhold.
Gehl, J. (2010). *Cities for People*. 1st ed. Island Press.
Gustafson, P. (2001b). Roots and Routes: Exploring the Relationship between Place Attachment and Mobility. *Environment and Behavior, 33*(5), 667–686.
Gustafson, P. (2009b). More Cosmopolitan, no less Local. *European Societies, 11*(1), 25–47.
Hartig, T. (2007). Three Steps to Understanding Restorative Environments as Health Resource. Cited in: Hartig, T. (2007). Open Space: People Space (1st ed). Taylor and Francis.
Haynes, R. (2008). Impact of Workplace Connectivity on Office Productivity. *Journal of Corporate Real Estate, 10*(4), 286–302.
Klinenberg, E. (2018). *Palaces for the People: How Social Infrastructure Can Help Fight Inequality, Polarization and the Decline of Civic Life*. 1st ed. Penguin.
Latham, A., & Layton, J. (2019). Social Infrastructure and the Pubic Life of Cities: Studying Urban Sociality and Public Spaces. *Geography Compass, 13*(7), 1–15.
Lewicka, M. (2013). Localism and Activity as Two Dimensions of People-Place Bonding: The Role of Cultural Capital. *Journal of Environmental Psychology, 36*, 43–53.
Lynch, K. (1960). *The Image of the City*. 1st ed. MIT Press.
Lynch, K. (1981). *A Theory of Good City Form*. 1st ed. MIT Press.
Lynch, K. (1984). *Good City Form*. 1st ed. The MIT Press.
Manzo, L. C. (2005). For Better or Worse: Exploring Multiple Dimensions of Place Meaning. *Journal of Environmental Psychology, 25*(1), 67–86.
Manzo, L. C. (2014a). Exploring the Shadow Side: Place Attachment in the Context of Stigma, Displacement and Social Housing. Cited In: Manzo, L

and Devine-Wright, P. (2014). *Place Attachment: Advances in Theory, Methods and Applications* (pp.178–190).1st ed. Routledge.

Massey, D. (1994). *A Global Sense of Place*. Cited in: Massey, D. (1994). *Space, Place and Gender* (1st ed.). Blackwell.

Masso, A. D. (2019). Between Fixities and Flows: Navigating Place Attachments in an Increasingly Mobile World. *Journal of Environmental Psychology, 61*(2019), 125–133.

McIntyre, N., Williams, D. R., & McHugh, K. E. (2006). *Multiple Dwelling and Tourism: Negotiating Place, Home and Identity*. 1st ed. CABI.

Murrain, P. (1993). *Urban Expansion: Look Back and Learn*. Cited in: Hayward, R., & McGlynn. (1993). *Making Better Places: Urban Design Now* (pp. 83–94). Butterworth-Heinemann.

Najafi, T. F., et al. (2016). Observation in Grounded Theory and Ethnography: What are the Differences? *Iranian Red Crescent Medical Journal., 18*(11), 40786.

Newman, O. (1973). *Defensible Space* (1st ed.). Architectural Press.

Stedman, R. C. (2003). Is It Really Just a Social Construction? The Contribution of the Physical Environment to Sense of Place. *Society and Natural Resources, 16*(8), 671–685.

Sun, S. S., Che, M. Z. & Liu, M. (2017). Research on Planning and Design of Community Pedestrian Spaces: Taking Tianjin Eco-City as an Example (生态城市规划初探——以中新天津生态城总体规划为例). Proceedings of the Urban Development and Planning Conference. Haikou, China. 1(1): 1–6.

Szerszynski, B. & Urry, J. (2006). Visuality, Mobility and the Cosmopolitan: Inhabiting the World from Afar. *British Journal of Sociology, 57*(1), 113–131.

Talen, E. (2002). The Social Goal of New Urbanism. *Housing Policy Debate, 13*(1), 165–187.

Urry, J. (2000). Mobile Sociology. *British Journal of Sociology, 51*(1), 185–203.

Williams, D. R. (2014). Making Sense of Place: Reflections on Pluralism and Positionality in Place Research. *Landscape and Urban Planning, 131*, 74–82.

Williams, D. R., & Kaltenborn, B. P. (1999). Leisure Places and Modernity: The Use and Meaning of Recreational Cottages in Norway and the USA. Cited In: Crouch. D. (1999). *Leisure Practices and Geographic Knowledge* (pp. 214–230). Routledge.

Williams, D. R., & McIntyre, N. (2001). Where Heart and Home Reside: Changing Constructions of Place and Identity. Trends 2000: Shaping the Future. Lansing (pp. 392–403). Michigan State University.

Williams, D. R., & McIntyre, N. (2012). Place Affinities, Lifestyle Mobilities and Quality of Life. Cited in Williams, D. R. and McIntyre, N. (2012). *Handbook of Tourism and Quality of Life Research* (pp. 209–231). 1st ed. Springer.

Williams, D. R., & van Patten, S. R. (2006). Home and Away? Creating Identities and Sustaining Places in a Multi-Centred World. Cited In: I.

N. McIntyre, & D. W. and K. McHugh. *Multiple Dwelling and Tourism: Negotiating Place, Home and Identity* (pp.32–50). 1st ed. CAB International.
Yang, B. J. and Dong, K. (2008). Theories and Practices of Eco-City Planning with Master Plan of Sino-Singapore Eco-City in Tianjin as an Example. *City Planning Review, 32*(8), 245–251.

CHAPTER 6

# The Relationship Between People and Place from the Perspective of Planning and Design Process

## 6.1 Constraints on Promoting the Behavioural Connection Between People and Place

In the strong top-down approach which is adopted during the development process of SSTEC, policies and law enforcement are two major drivers of meeting quantitative efficiency targets in the building of the eco-city. Nevertheless, public's requirements and acceptance of the planning project are the last factors to be considered (Hu et al., 2015, p. 6). The challenge faced by the previous planning and later management stages of the SSTEC tends to focus on the flexible adjustment of design guideline—people's (esp. residents') behaviours are diverse and their needs are dynamic, additional new design requirements cause the increase of management cost, which has been seen as the main obvious obstacle for realizing the people-oriented concept in the eco-city project. Evidences that support such argument will be further discussed in the following section by using the thematic analysis for the transcript of interviews as the first-hand data as well as the document analysis for some representative academic sources as the second-hand data.

As shown in Chapter 4, during the initial period of eco-city's construction, national governments of both China and Singapore jointly formulated an overall regulatory indicator system, covering twenty-two controlling indicators and four guiding indicators within social, economic and



Y. Wu, *Possibilities and Limitations of Chinese Eco-City Development*, New Perspectives on Chinese Politics and Society,
https://doi.org/10.1007/978-981-95-1958-3_6

environmental aspects. These indicators have been regarded as quantitative goals for eco-city construction and bases for guiding planning and management. At the level of the regulatory detailed plan, clear requirements have been put forward on the land use, scale and supporting social infrastructure of public space at both urban and community level, such as eco-valley, waterfront areas, community park, etc. However, strict requirements of the street-level public space, such as slow-traffic system and pocket park, have not been clearly defined regarding their land-use pattern, scale and locations in the eco-city's regulatory detailed plan. As a consequence, public space at the street-level cannot be effectively controlled during the regulatory planning phase. Moreover, requirements of humanistic planning and design need flexible adjustment and modification during the implementation process because people's needs are diverse and can change over time, it has made the eco-city management sometime unresponsive to those changing situations in a timely manner.

> Specifically, guidance for space-making by urban design alone cannot form an effective planning legal basis, as the eco-city project has given property developers too many permissions to change scale, locations and supporting social infrastructure in the street-level public space. Thus, in terms of the public space, the original intention of good people–place relationship or people-oriented principle could not be effectively implemented at the construction phase, making the integrity of public space system within the whole eco-city be impacted at the initial planning stage. (Member of Tianjin Eco-City Administrative Committee, interviewed on 20 July 2019)

> As you (researcher) mentioned about the 3rd community centre of SSTEC, regulatory detailed plan requires to reserve (预留) convenient social facilities in the community construction, such as enough resting area and barrier-free facilities. Although planning approval and construction process have done in accordance with this requirement, due to some management factors in the post-operational period, such as high operating cost for the replacement has led the former-designed social facilities and former-planned community services could not be implemented at last. Also, additional new planning and design requirements, such as more municipal-supporting social facilities should be built according to residents' needs by some public participation methods, have been told to planning practitioners during the constructional period, or even during the period that social facilities have been put into use. Such situation could be seen as an emergency (突发事件), and even if issue can be solved eventually, it will

still cause the increase of management cost'. (Member of Tianjin Eco-City Administrative Committee, interviewed on 20 July 2019)

As the transcript of above interview shows, although planning approval and construction process have been in accordance with this requirement, due to some management factors in the post-operational period, such as high operating cost for the replacement, the former-designed social facilities and former-planned community services could not be implemented at last. Also, as people's needs are dynamic and change over time, it is difficult to fully meet people's needs in a short period of time, including their ever-growing psychological needs (esp. sense of safety when staying in the public space discussed in the previous chapter), and even their basic needs about social infrastructures (e.g. more barrier-free facilities and more resting area in the public space discussed in the previous chapter). As the transcript of above interview shows, planning approval and construction process have been in accordance with this requirement, due to some management factors in the post-operational period, such as high operating cost for the replacement, the former-designed social facilities and former-planned community services could not be implemented at last. Such constraint has also been discussed in the paper named *Study on the Implementation and Evaluation of the Regulatory Plan in Zone Five of the SSTEC*—as Lu et al. (2019) pointed out, two main flexible problems for eco-city construction in terms of the financial issue all remain at the construction stage. On the one hand, the eco-city's construction of social infrastructures (public utilities) is mostly monopolized by the investment company, thus, eco-city's construction and development could be slow and costly without a healthy market competition mechanism. On the other hand, the eco-city aims to provide people the healthy and liveable environment (Bibri, 2020, p. 7); high standards and high requirements for social infrastructures have brought about high expenditure and high time-consumption. Under such circumstances, in terms of social infrastructure, the contradiction between the quality of urban construction and funds cannot be resolved and can only be alleviated by delaying expenditures, consequently, residents' ever-changing needs could not be met timely.

The above-mentioned issue cannot be completely resolved by only redeveloping or reconstructing specific public space as it indeed causes a great deal of manpower and financial resources. As the previous chapter reveals, social infrastructure is regarded as means of realization for the interactive connection between place, Behrens and Watson (1996) recommended, social infrastructures or public facilities, which are functionally related should be located in clusters in the face of limited public funds (p. 83). Faced with such a dilemma, planning practitioners, especially urban designers, may be able to try a more active response, such as in the previous design stage to set aside more flexible process and to avoid the unbearable operating costs in the later management stage.

This research starts from the perspective of residents to discuss the conceptual-practical gap of humanistic concept in the eco-city development, then it uses the perspective of planning system to explore reasons about causing such constraints. After combining these two aspects as a whole for analysis, it is obvious that planners or designers of the eco-city are those who can communicate directly with residents to obtain the first-hand information, such as understanding people's needs and capturing their behavioural preferences. Thus, discussion about the implication of people–place interactive relationship has been considered from the perspective of design procedure.

## 6.2 Information Interpretation Matters During the Design Process: The Importance of Leverage Points of the Place

'Attempts to develop users' requirements as a basis for physical design in building reflected the desire for a more rational approach in the light of the increasing complexity of problems involved in design and the increasing separation between designers and users' (Rubin & Elder, 1980, p. 83). It is clear from even the most cursory glance at manifestos of modern famed designers that a desire to create an environment full of opportunities and availabilities for a liveable life is deeply rooted in planning philosophy. For example, 'Le Corbusier made the point that the ultimate measure for buildings must be man' (Rubin & Elder, 1980, p. 92). However, in reality, the tremendous gap between designers' intentions and their achievements indeed exists—'this disparity between success and failure illustrates the difficulty that designers do not have the capacity to predict the outcome of designs, ... [namely] the impact on people with any reasonable probability of accuracy' (Lang et al., 1974, p. 3). Specifically, designers are often heard to ask, 'why are people so stubborn or misguided as not to use the place and spaces we design—either not at all or in the wrong way?' (Perin, 1970, p. 23) Such design issues indeed happen in SSTEC. For example, the researcher has observed an interesting phenomenon—some children tend to enjoy playing with clay at the entrance of children's park, instead of staying in the children's park for playing. As discussed in the previous chapter, different age group's needs are different in public spaces which demands understanding about how different age group interacts with environment. From this perspective, people's behaviours within the environment can be explained by the theory of *Environmental Psychology*, as it is a field that focuses on understanding how individuals interact with and how they are influenced by the physical environment. Take behaviour setting theory as an example, it focuses on the role of specific environment in influencing consistent patterns of behaviour, and different environment creates expectations for appropriate behaviour. It is important to note that people's behaviours are often influenced by a combination of factors, and *Environmental Psychology* theory helps to understand the complex interplay between individuals and their surroundings.

As Rubin and Elder (1980) demonstrated, 'we need to upgrade the quality and importance of users' information employed in the design process. If [places] are to serve people who [use] them, we must learn more about the way in which the built environment affects behaviours and about how behavioural factors should influence design' (p. 4). From this perspective, people (users) act as design participants, their information shows how attributes of place and behaviours of people interact. 'The behavioural sciences [rooted in environmental psychology] can help us to understand the present and what the trends in society are, they can help us to predict outcomes of our design proposals for the future better than we do now' (Lang, 1987, p. 29). Therefore, it is possible to reduce the *performance uncertainty* for not only designers but also urban planners while making decisions on users' needs. As Deasy and Lasswell (1985) pointed out, 'useful information must be located, and it must be translated from the terminology of human science to the language of design' (p. 11). In this research, people's needs or their behaviours are two important aspects for information interpretation in the eco-city's design process.

In terms of eco-city's planning practices, creating pedestrian-friendly and mixed-use public spaces has always been the development goal of the SSTEC. In recent years, the SSTEC has improved the quality of urban public spaces by connecting greenways and gathering residents' daily lives. Since the release of the latest indicator system (2.0 version), eco-city's planning and design cares more about people's preferences and their needs. When asked about how people are regarded as an important element during eco-city's design process, namely how people's needs are reflected in planning and design methods, *the Deputy Director of Sino-Singapore Tianjin Eco-City Construction Bureau* illustrated two main aspects, public spaces and the 4th community centre put into use in the year of 2022 (Fig. 6.1).

> Since the release of the latest indicator system (2.0 version), the SSTEC has focused on two key indicators, route accessibility and green landscape respectively, to better utilize the function of public space and to realize the people-centred design goal. In 2022, the implementation of the Ministry of Housing and Urban–Rural Development started to carry urban physical examination work(城市体检工作),the SSTEC conducted a comprehensive assessment of the existing public spaces (especially the greenway system), and categorized into three levels – excellent, good and inadequate. Based on the current state of public spaces, with the goal of serving residents and improving their living environment, the SSTEC has closely followed the concept of green and liveable urban construction, and also has determined

**Fig. 6.1** Digital information presented on the screen of the operation centre in real time (*Source* Figure is taken by the researcher)

the first-phase greenway and community parks planning. In addition, under the Digital China Strategy (数字中国战略), the SSTEC plans to accelerate the integration of smart city and low-carbon city construction by carrying out intelligent upgrading of the greenway system - through intelligent parking, intelligent lighting, environmental monitoring, greenway system has been provided better services for citizens to travel. (Deputy Director of Sino-Singapore Tianjin Eco-City Construction Bureau, interviewed on 16 May 2023)

Starting from the industrial development and residents' needs, the SSTEC takes both ecological and smart elements into full consideration. The eco-city adheres to people-oriented and application-first planning concepts in order to build a smart eco-city with leading value. As you know, the central area of the eco-city is the construction pilot of Complete Residential Community(完整居住社区), with the area of 4.4 square kilometres. The first project launched in the pilot area is the 4$^{th}$ community centre, which aims to create a community that is safe, energy-saving and ecologically-vibrant. Breaking through the rigid division of land use boundaries, and based on the evaluation of results of urban physical examination work(城市体检工作), the 4$^{th}$ community centre uses design approaches that take people's needs and feelings as the primary criterion to improve spatial functions. It integrates community centres, community parks, primary schools and kindergartens within one neighbourhood to facilitate people's daily lives, especially for the surrounding residents. (Deputy Director of Sino-Singapore Tianjin Eco-City Construction Bureau, interviewed on 16 May 2023)

### *6.2.1 People's Needs to Measure Their Behaviours*

After finishing the fieldwork, the main reflection for the researcher is about how eco-city can achieve people-oriented planning goal to the maximum extent, especially how eco-city's planning and design can encourage people to do more outdoor activities and make people feel safe, comfortable and delighted when staying in public spaces. According to the *New-Type Conceptual Model of Human-Centred Design*, organizing people's needs on the basis of low-to-high level has been seen as a good way to discover the value of a place. Alfonzo (2005) examined people's activity needs of the linear public space (walking network). By analysing the psychological process about whether people decide to walk or not, Alfonzo sorted different factors that can affect this process by their importance in a hierarchical relationship (*See* Fig. 6.2).

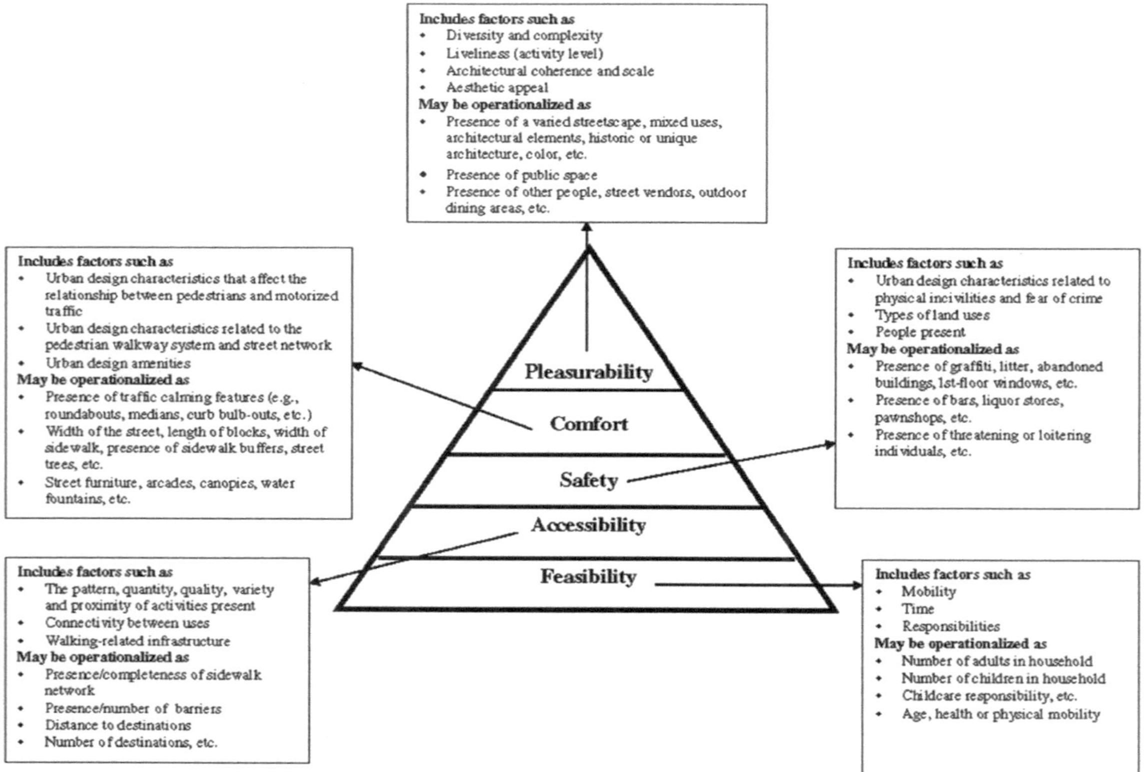


**Fig. 6.2** Hierarchy of walking needs (*Source* Figure is from Alfonzo, 2005, p. 825)

Based on Maslow's theory, Alfonzo believes that a person who is at a lower need level could not be aware of the higher-level needs. When a series of needs cannot be met, the bottom-level ones should be the first priority. Sense of safety (psychological aspect) and social infrastructure (physical aspect) have been identified as two forms of behavioural connection between people and place, and also been deemed as two important *leverage points* of eco-city's place-making process. This viewpoint can help planners or designers better understand issues and discover the *leverage points* (effective operative goal or *pain point* of the place) during the eco-city's place-making process. For example, data analysis of the Zhongxin Avenue shows place lacks protective features of meeting residents' needs for safety, so the pursuit of safety features should be seen as the *leverage points* of this site.

Eco-city planning aims to take the people-oriented concept as the main design goal, which has required designers to grasp users' needs. A number of theoretical discussions have paid attention to the question about what are people's needs in an academic context, and the most widely known is the *Maslow's Hierarchy of Needs* mentioned. However, the obvious limitation about introducing this theory into urban design discipline has been seen as it generally describes needs for people's life, a certain deviation and unique interpretation about people's specific needs towards the physical environment indeed exists, such as the senses of safety, comfort, enjoyment, as well as satisfaction about social infrastructure. Therefore, it cannot completely copy the theoretical description of *Maslow's Hierarchy of Needs* into eco-city's planning process. Academic theories, especially Maslow's Hierarchy of Needs mentioned in this research can help planners to deeply understand people's universal needs, however, it is not enough to stay at the theoretical level for taking people's needs into account.

Popularize from Zhongxin Avenue to the overall eco-city, seeking for the *leverage points* of a specific space is not only appropriate for the design work (e.g. pedestrian network), but also promotes other types of eco-city's spatial design (e.g. resting space optimization). Specifically, during the place-making process, planners discover key points of people's needs that the site does not meet (deficiency). Afterwards, planners organize them in an orderly manner based upon the hierarchical model—during the redevelopment phase, the priority task should be seen as meeting people's lower-level needs in order to make the most efficient use of the limited construction fund. As Whyte pointed out, 'supply creates demand.

A good new space builds a new constituency. It stimulates people into new habits' (p. 16). Jan Gehl (2010) had a similar viewpoint, 'potential human needs can be stimulated when supported by suitable material conditions' (p. 41). From this point of view, people-oriented design means identifying *leverage points* of the place, which can be caught by the behavioural connection between people and place. Moreover, two forms of such behavioural connection—sense of safety (psychological aspect) and social infrastructure (physical aspect), should be considered as two important *leverage points* of the space, in order to realize the harmonious people–place interactive relationship in the eco-city project.

The fieldwork of the SSTEC has provided evidences to support the hierarchy needs model. As research data shows, interviewees are more concerned about the lower-level basic needs (e.g. easily accessible resting areas and social infrastructures, safety of public space and convenience of transportation) rather than higher ones (e.g. aesthetic or eco-technical requirements). In this research, it mainly focuses on the measurement of people's needs survey. Chinese eco-city planning takes the people-oriented concept as the main planning goal and design target, which has required planning practitioners to grasp users' needs in different procedures, such as the previous planning and later management stages.

Even in the same research location, satisfaction about people's needs has often been a combination of behavioural types, which makes people's needs more likely to be invisible. Take 3rd community centre and 4th community centre as examples. Although it is used for people's grocery shopping, and also the majority of people (interviewees) mentioned *grocery shopping* is their main motivation for coming here, social contact (e.g. chatting while waiting) becomes the equally important purpose based on researcher's observation records. As a result, comfortable resting environment and the pleasure people feel in this place have become people's invisible needs, which sometimes cannot be directly reflected from people's words.

### 6.2.2 *Judge the Differences*

*Garden City* and *Ecological Urbanism* theories have portrayed what a harmonious people–place urban form should be. Referring to the *Garden City* theory, it is of significance to rethink the *Garden City* concept for the twenty-first century, and in particular to rediscover its benefits about delivering attractive, liveable and sustainable new communities—'the breakthrough and more comprehensive response to the grim, environmentally degraded industrial cities symbolically associated with the grid forms' (Dobbins, 2009, p. 40). Howard's transcendent aim was to create an urban settlement that would be better for everyone's living there, not just the landowner and the developer. His garden city proposals in 'diagrammatic abstractions projected a new way for conceptualizing cities, which picked up from utopian notions of embodying both physical and social content as the drivers for new forms' (*ibid*). Referring to the *Ecological Urbanism*, this theory intends to make this emphasis as far as possible—it overlaps *Urban Ecology* when it considers ecosystem as a syncretic element of human and natural. The focus on *Ecological Urbanism*, however, is on the urban aspect, rather than the ecosystem. At the conceptual level, *Ecological Urbanism* is 'a cultural practice that demands from the built environment, giving a change in thinking' (Hagan, 2015, p. 23).

Transition from the conceptual advocacy to the practical implementation, information interpretation requires eco-city's designers to act like doctors, as *Ecological Urbanism* advocates, making the 'diagnoses' for both built environment and social infrastructures in order to judge the 'gap' between current situation of public space and the ideal built environment they hope to design during the place-making process. In the place-making process, designers can judge the 'gap' between current situation of public space and the ideal physical environment by two ways. One is to compare differences between actual spatial characteristics of public spaces and high-quality spatial characteristics, from the perspective of both spatial elements and spatial environment. As the previous chapter demonstrates, this research defines high-quality public space within the eco-city as space with the feature of harmonious people–place relationship, namely eco-city's urban design not only meet people's lower-level needs (e.g. safety) when they use the public space to do outdoor activities, but also meet their higher-level ones (e.g. enjoyment). However, characteristics of high-quality public space sometimes could be difficult to quantify, consequently, it is not easy to make judgements about the gap. Jacobs and Appleyard (2007) attempted to summarize elements of good environment through a large number of examples.

> A number of goals that we deem essential for the future of a good urban environment: liveability, identity and control, access to opportunity, imagination and joy, authenticity and meaning, open communities and public life, self-reliance and justice. ... The good urban environment is one that somehow balances these goals, allowing individual and group identity while maintaining a public concern, encouraging pleasure while maintaining responsibility, remaining open to outsiders while sustaining a strong sense of localism. (Jacobs & Appleyard, 2007: 115–116)

Due to unique characteristics of each locality, making this people–place harmonious concept be fully clarified or forming a unified standard seem to be tricky. Those measurable requirements are rarely accurate to a certain quantitative value, such as the number of resting seats that people think enough and the number of street light people feel safe. Thus, this comparative measurement for designers to judge the gap between current situation of public space and the ideal physical environment they plan to design will be a range as some spatial concepts do not have strict and unified good or bad quality standards.

Another way is to examine whether the built environment is compatible with the expected planners' and designers' intentions, from the perspective of users' behavioural patterns. Kevin Lynch in his book *Good City Form (1984)* demonstrated the degree of good city performance was determined by its ability to provide ecological, psychological, social and cultural requirements to its inhabitants. Fit, one of five performance dimensions to measure good physical environment, refers to the match between people's behaviours (function) and the physical environment (form). It is 'how well the spatial and temporal pattern of a settlement matches the customary behaviour of its inhabitants' (Lynch, 1984, p. 151). As data conclusion shows the value of behaviours, when there is congruence between physical urban form and patterns of people's behaviours, people feel comfortable and delight. Conversely, absence or lack of *fit* can make people feel uncomfortable and difficult to behave in the public space.

In short, the reflection of people-oriented concept in the practical project is seen as positive (harmonious) behavioural connection between people and place. Even though the relationship between space and behaviour is not decisive, it is not easy to judge whether there is a *gap* between the planned public space and the expected behavioural pattern. Nevertheless, *fit* is not a rigid application of spatial standards used in previous planning projects. It should be creative and flexible for urban designers to improve their foresight abilities about the connective relationship between physical environment and people's behaviours by their thinking and practices.

## 6.3 The Existing Traditional Eco-City's Design Process

One key point that needs to be firstly highlighted is the process about translating eco-city's KPIs to the statutory plan. 'The vision of eco-cities can become reality through implementation' (Yip, 2008, p. 3). Thus, to take the concept to planning practice, it is necessary to integrate planning concept and objectives into the current statutory planning system. Following the traditional Chinese planning and design process, in order to maintain the integrity of KPI performance at detailed planning implementation level, Chinese eco-city project has undertaken to translate the masterplan concepts and KPIs into detailed regulatory plans. The framework is to translate the planning objective guided by the KPIs into usable land-use standards as well as urban design and building design codes as part of the statutory planning management system (Fig. 6.3).

Key point that needs to be generally highlighted is about the main weakness referring to the planning and design process of Chinese eco-city. As Yip (2008) pointed out, the traditional planning system has put great focus on setting out site-specific development parameters, such as density and plot rations at the local detailed plan level, namely the statutory regulatory plan.

> It is at the regulatory plan level that the specific development and planning constraints are laid down on the plan as part of the legal requirements. These planning conditions based on the regulatory plan are also included into land use right lease agreement between the government and the land owners/developers. However, the current mandatory planning parameters used in the regulatory plan are limited in scope and may not be able to drive all the eco-city planning intentions at the site-by-site level. (Yip, 2008, p. 5)

Guidance for space-making by urban design alone cannot form an effective planning legal basis, as the eco-city project has given property developers too many permissions to change scale, locations and supporting

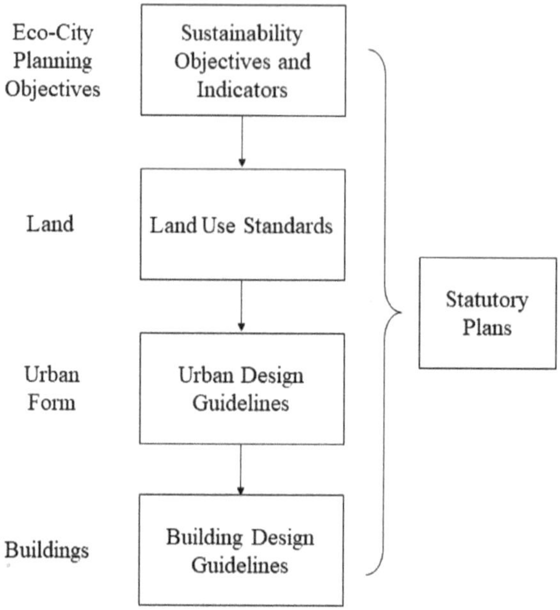


**Fig. 6.3** Translating eco-cities' KPIs to statutory plans (*Source* Figure is drawn by the researcher)

social infrastructure in the street-level public space. Consequently, the original intention of people-oriented principle could not be effectively implemented at the construction phase, making the integrity of public space system within the whole eco-city be impacted at the initial planning stage. Such general weaknesses of Chinese eco-city planning and design system may respond to the question about why there are two good-quality places and two poor-quality places within the selected four research locations, which are all operated under the same process of eco-city's planning and design. In response to this question, the explanation has been given by the *Deputy Director of Sino-Singapore Tianjin Eco-City Construction Bureau,* Peng Fu.

> In some cases, property developers have been granted a certain degree of influence and permissions in the SSTEC planning projects, which can lead to concerns about the potential for developers to change the scale or to alter original intentions of these eco-city projects. There are three

> main reasons why this situation might happen. Firstly, due to the economic pressure. Property developers often have financial interests in maximizing their returns on investment. They might seek to increase the scale of the development to accommodate more housing or commercial space, which can generate higher profits. Secondly, due to the market demand. Developers might argue that there is a strong market demand for larger or different types of properties within the eco-city. This could lead to the request for changes that align more with market preferences, potentially diverging from the original eco-friendly and sustainable goals. Thirdly, due to the regulatory flexibility. In some cases, local governments might provide developers with flexibility in adhering to certain regulations or guidelines to encourage economic development. This could inadvertently allow developers to make changes that prioritize profit over eco-friendly features or people-oriented concept. (Deputy Director of Sino-Singapore Tianjin Eco-City Construction Bureau, interviewed on 16 May 2023)

Dividing the object of eco-city's place-making into three categories, people's behavioural activities, people's preference and needs and also physical environment elements, has formed a one-to-one correspondence with the model *Sense of Place* proposed by John Punter (Fig. 6.4).

As the literature review chapter demonstrates, designing liveable place for people deems as the core aim of eco-city planning, and creating the sense of place for people not only depends on the form of physical environment, but also relies on meaning and people's activity, other

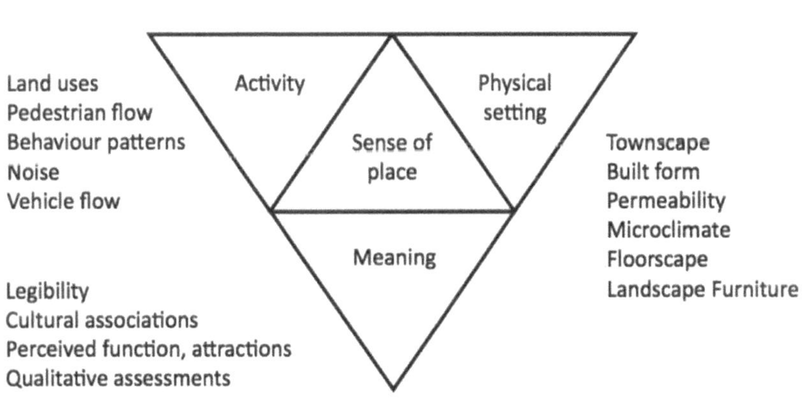


**Fig. 6.4** Components of sense of place (*Source* Figure is from Punter, 1991, p. 27)

two equally important elements. And as the previous chapter illustrates, since the revised version of indicator index has shifted from traditional economy-oriented pattern to current people-oriented pattern, urban planners need to ensure the designed places of the eco-city can better meet people's needs and form a better activity mode by maximally understanding users' cognition, behaviours and elements of the existing physical environment. In this sense, the process of eco-city's place-making can be regarded as the procedure of obtaining an *upgraded mirror image (See* Fig. 6.5*)* composed by the sense of place.

Based upon the interview with the official of *Tianjin Planning Bureau*, the design process of eco-city's place-making has been divided into four stages, including programming, schematic design, implementation and evaluation.

> Firstly, the initial stage is to collect information which relevant to specific eco-city's design situation, known as programming. Secondly, the core stage of the eco-city's place-making, refers to the schematic design. Designers analyse the information collected in the programming phase to put forward ideas, such as preliminary space allocation plans or design details. At this stage, successful schematic design plan needs to provide

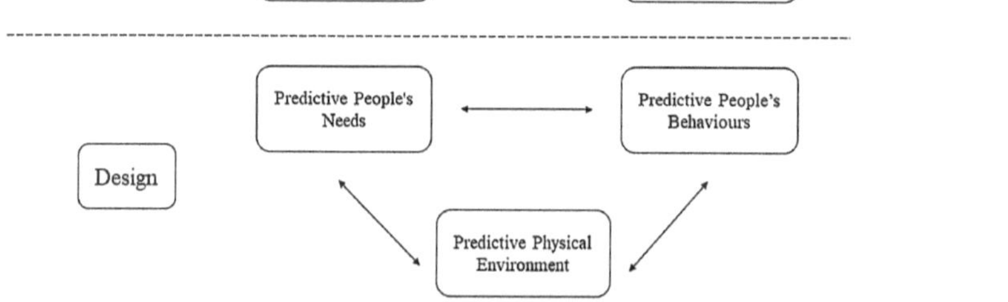


**Fig. 6.5** Place-making considering the people–place relationship (*Source* Figure is drawn by the researcher)

clear picture of eco-city's design direction or characterises of the built environment. Thirdly, implementation stage involves the production of design proposals, providing detailed information for eco-city construction. Lastly, evaluation stage investigates how effective eco-city's design service residents and how well eco-city's design meets its stated goals. Even though this is a follow-up stage of the completed design work, it is vital for the entire process as 'inefficient performance' can be noted for further design improvements. (Official of Tianjin Planning Bureau, Member of SSTEC Group, interviewed online on 10 March 2021)

Planners find out the dominant problems and opportunities of the site, and to be honest, ranking people's needs from the lower-level to the higher-level or their behaviours when using the public space, as you mentioned some examples, have not been examined in SSTEC's planning process. (Official of Tianjin Planning Bureau, Member of SSTEC Group, interviewed online on 10 March 2021)

Based on the transcript of interviews, in the traditional eco-city's place-making process, urban design of the eco-city can be roughly divided into four steps, shown as follows.

Fieldwork based on the spatial analysis, which is supplemented by planners' participation observation or on-site interviews.

Synthesize information collected in the first step to find out threats and opportunities (often use the *SWOT Analysis Method*) as well as to extract design goals or quantitative indicators.

The process of designing conception. At this stage, it may require additional supplementary fieldwork to double-check site issues found in the previous research.

The formulation of the final design proposal (Fig. 6.6).

## 6.4 Recommendation for Practice

### *6.4.1 Possible Improved Design Process Considered About the People–Place Relationship*

As the previous section *Constraints on Promoting the Behavioural Connection between People and Place* discusses, the challenge faced by the SSTEC tends to focus on the flexible adjustment of design guideline—people's (esp. eco-city's residents) behaviours are diverse and their needs are dynamic, additional new design requirements cause the increase of management cost, which has been seen as the main obvious obstacle for

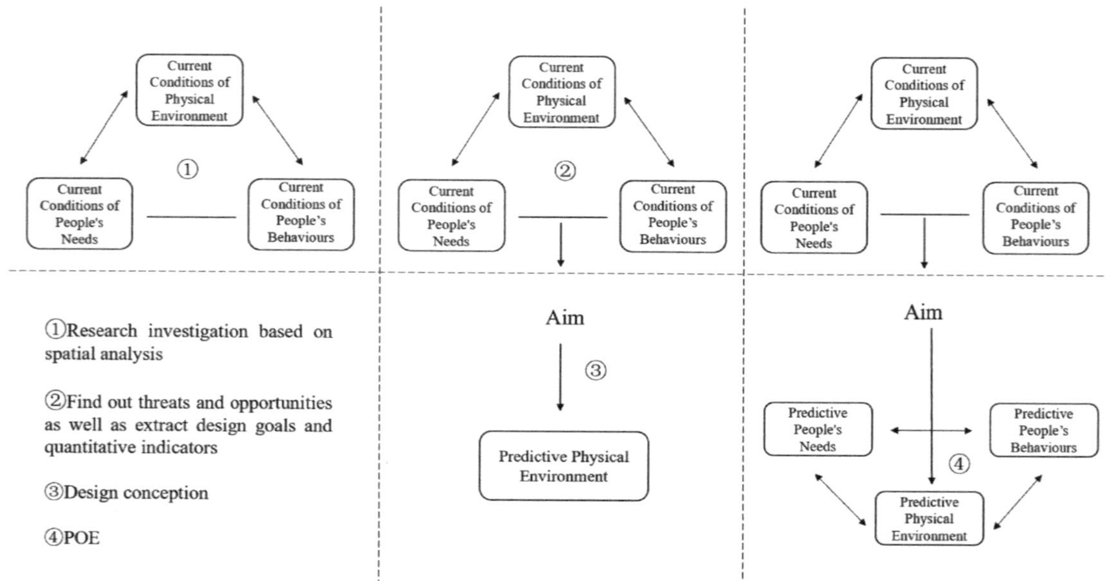


**Fig. 6.6** The traditional eco-city's place-making process (*Source* Figure is drawn by the researcher)

realizing the people-oriented concept in the eco-city project. For example, people's psychological feeling (e.g. sense of safety) has not been fully considered at the design and use level, and people's using preferences of social infrastructures (e.g. sufficient resting area, convenient and friendly facilities for the vulnerable people) are not enough predicated, in the later operational and management stages, additional manpower or financial resources have to be invested for eco-city's redevelopment to meet the changing people's needs. This section aims to recommend one possible improved design process considered about people–place relationship, in order to tackle the flexible adjustment of design guideline.

Much effort has been put to model the process of urban designing by the academia. While explicit models can give designers a clear structure, urban design does not always take place in the neat sequential manner that models suggest. It is a 'highly argumentative process of conjecturing—putting out ideas—and testing them in an iterative fashion' (Lang, 2005, p. 26). Is it possible to analyse the rational derivative process that occurs in the mind of designers in order to clarify this process? *Meeting people's needs* acts as one of the primary eco-city's planning and design goals, to realize it rather than make it become a slogan, it is necessary to make some modifications or improvements for the traditional design model. Specifically, the implication of people–place interactive relationship focuses on predicting and integrating *people's cognitive and behavioural pattern* into other place-making procedures, to make the formation process of design concepts become clearer and much human-centred (Fig. 6.7). As Fig. 6.8 shows, the traditional eco-city's place-making process has been modified from four to six steps. This possible improved design process is inspired by the model of rational design process (*See* Fig. 6.7)—the process of urban design is 'a high argumentative process of conjecturing, putting out ideas and testing them in an interactive fashion' (Lang, 2005, p. 26).

In this rational model, urban design has been seen as 'an argumentative process in which participants in it learn as they go along' (Lang, 2005, p. 27). Designers learn about goals, design methods and evidences as perceived by other stakeholders. They argue about variables which ought to be taken into consideration and what good design entails. Last but not least, 'conjectures are tested by individuals using their own logics based on their predictions of the consequences of different design actions' (*ibid.*). In such process, designers learn about goals and means as perceived by stakeholders, they argue about variables that should be taken into full consideration and what good design entails. This rational model

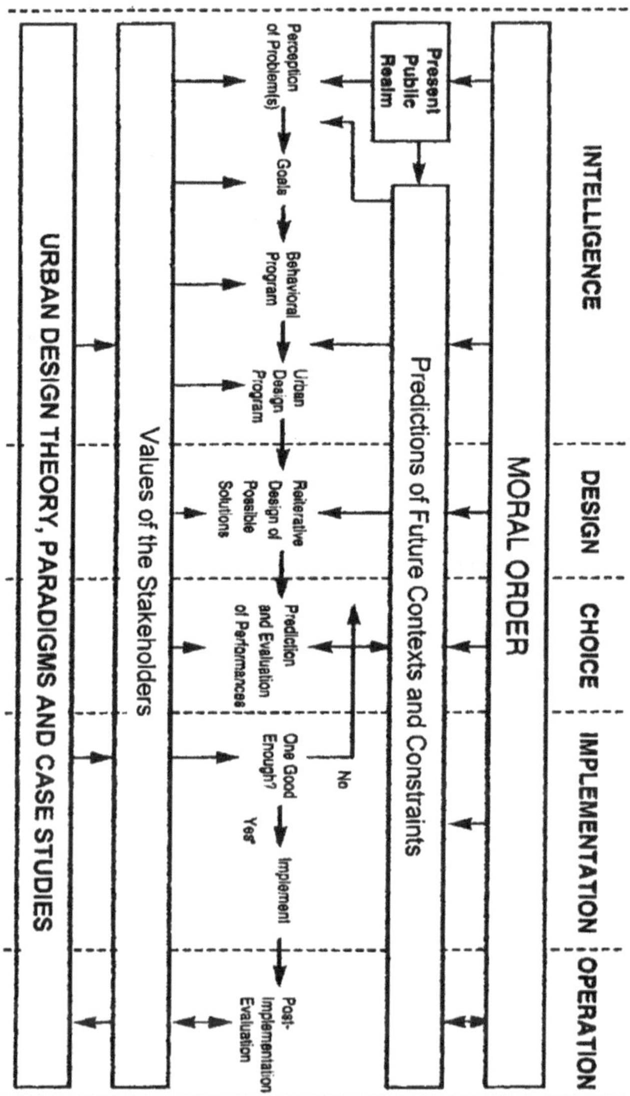


**Fig. 6.7** The model of the rational design process (*Source* Figure is from Lang, 2005. p. 26)

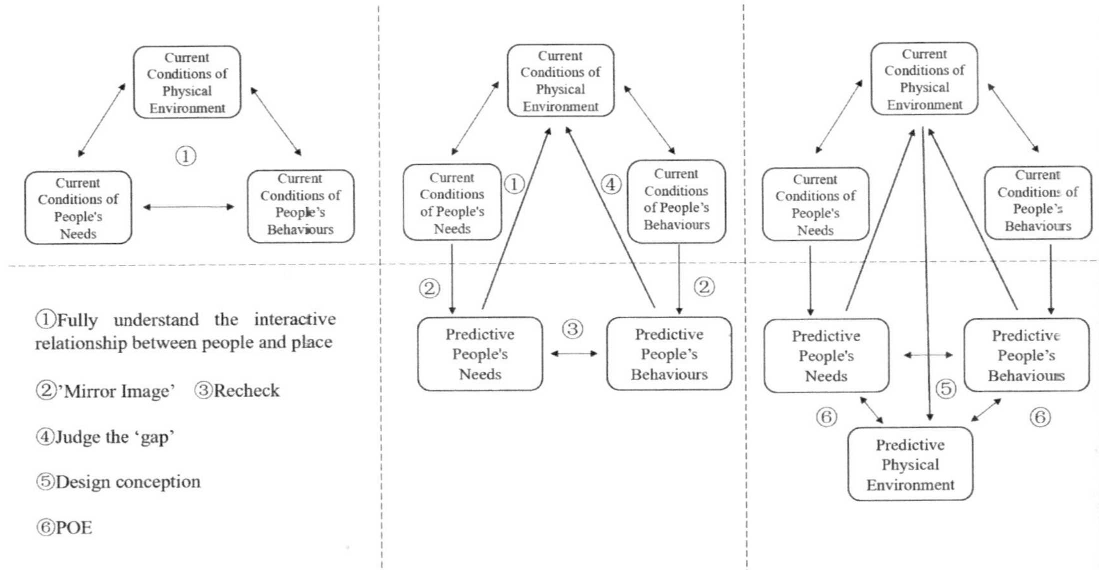


**Fig. 6.8** The improved eco-city's place-making process (*Source* Figure is drawn by the researcher. May 2021)

suggests that the process of urban design begins with the perception of opportunities worth exploring or problems worth overcoming.

Compared with the process of traditional eco-city planning model, this modified version has added a requirement for planners to consciously create the *upgraded mirror image* referring to the three-level sense of place, and has reduced the procedure of summarizing design goals and quantitative indicators. That is, by actively seeking for the *gap*, more attention about people–place interactive relationship, which takes the behavioural connection into consideration, has been paid in this improved model, which reflects the future's scene that people use public space in the eco-city.

- Do the site fieldwork to fully understand the relationship between users' activities/cognition and the characteristic of physical environment.
- Conceive people's potential behaviours when they stay in the built environment for the future scenarios on the basis of current patterns of their cognition and activities. And check the relationship between expected people's cognition or activities and the subsistent characteristic of built environment.
- Designers need to objectively judge the 'gap' between the expected situation and the subsistent environment to form the detailed design proposal.
- POE verification work should be done after eco-city's construction where possible.

### 6.4.2 *Eco-City Planning Practices to Respond to Concept of Humanistic Planning Under the Idea of Ecology*

Incorporating people's needs and behaviours into Chinese eco-city's planning and design is not only a simple procedural addition or a strategic move to eco-city's design process, but also a necessary one that promises a host of substantial advantages. This research regards Chinese eco-city development represented by the SSTEC project as the corresponding product to adapt to the change of *New-Type Urbanization* development pattern—the gradual transformation from the newly built city construction which mainly focuses on economic and environmental development under the traditional urbanization pattern to the trinity of economic-environmental-humanistic urban governance, which is characterized by the concept of ***humanistic planning under the context of ecology*** based on the green urbanization development pattern. The improved eco-city's place-making process mentioned above adheres to a principle that creating the physical environment that caters to people's diverse behaviours as well as their own needs and preferences. First of all, integrating people's needs and behaviours into eco-city planning contributes to fostering the sense of place. By designing public spaces that encourage social interaction, such as children's park and community centre mentioned in this research, not only enhances people's quality of life but also encourages people's shared responsibility for the participation in place-making during the eco-city's later operational stage. In addition, resource efficiency is another key advantage. When urban planners align social infrastructures with daily routines of inhabitants, resources like energy and waste management could be optimized to match people's actual consumption pattern. This alignment, which people's behaviour pattern has been regarded as the evaluation criteria, could minimize waste and ensure essential resources are used judiciously. For instance, to meet the requirement of smart eco-city development, smart energy management systems can be integrated to harness renewable sources like solar power, efficiently meeting the energy demands of residents. In short, incorporating people's needs and their behaviours into Chinese eco-city planning and design is not merely an option but a critical strategy for building sustainable and vibrant urban environment. Ultimately, the eco-city designed with a deep understanding of people's needs and their behaviours is more likely to stand as a shining example of Chinese sustainable development, setting new standards for urban living in the green urbanization development pattern.

## 6.5 Conclusion

Humanistic planning emphasizes the well-being and people's quality of life as central point to eco-city planning and design. As the pilot eco-city project in China, the SSTEC has been confirmed to apply the people-oriented planning and design concept into its development process. However, implementation gap indeed exists; people still have different kinds of dissatisfaction when living in the eco-city, such as their physical needs towards social infrastructures and their psychological needs towards the environmental security. This chapter starts from the perspective of planning system to argue the traditional Chinese planning tends to be difficult to meet the dynamic needs of rapid urban development, which has resulted in the pressure about the flexible adjustments for design features, such as additional manpower or financial resources have to be invested for eco-city's redevelopment to meet the changing people's needs. Taking such constraints on implementation as the starting point, this chapter proposes the possible improved eco-city's design process, from the traditional four steps to six steps based upon predicting and integrating people's needs and their behavioural pattern, in order to mitigate limiting impacts on the conceptual-practical gap accompanying the realization of humanistic planning concept in the Chinese eco-city project by introducing the people–place interactive relationship.

## References


Alfonzo, M. A. (2005). To Walk or Not to Walk? The Hierarchy of Walking Needs. *Environment and Behaviour, 37*(6), 808–836.

Behrens, R., & Watson, V. (1996). *Making Urban Places: Principles and Guidelines for Layout Planning* (1st ed.). UCT Press.

Bibri, S. E. (2020). The Eco-City and Its Core Environmental Dimension of Sustainability: Green Energy Technologies and Their Integration with Data-Drive Smart Solutions. *Energy Informatics, 2020*(3), 1–26.

Deasy, C. M., & Lasswell, T. E. (1985). *Designing Places for People: A Handbook on Human Behaviour for Architects, Designers and Facility Managers* (1st ed.). The Whitney Library of Design.

Dobbins, M. (2009). *Urban Design and People*. 1st ed. John Wiley & Sons Inc.

Gehl, J. (2010). *Cities for People* (1st ed.). Island Press.

Hagan, S. (2015). *Ecological Urbanism: The Nature of the City* (1st ed.). Routledge.

Jacobs, A., & Appleyard, D. (2007). Toward an Urban Design Manifesto. *Journal of the American Planning Association, 53*(1), 112–120.

Lang, J., Burnette, C., Moleski, W., & Vachon, D. (1974). *Designing for Human Behaviour: Architecture and the Behavioural Sciences* (1st ed.). Wiley.

Lang, J. (2005). *Urban Design: A Typology of Procedures and Products* (1st ed.). Architectural Press.

Lynch, K. (1984). *Good City Form* (1st ed.). The MIT Press.

Perin, C. (1970). *With Man in Mind: An Interdisciplinary Prospectus for Environmental Design* (1st ed.). MIT Press.

Punter, J. (1991). Participation in the Design of Urban Space. *Landscape Design, 200*, 24–27.

Rubin, I., & Elder, J. (1980). *Building for People: Behavioural Research Approaches and Directions* (1st ed.). Supt. of Docs.

Whyte, H. W. (1980). *The Social Life of Small Urban Space* (1st ed.). Conservation Foundation.

Yip, S. C. T. (2008). Eco-city: From Concept to Implementation-Case Studies in Shanghai Dongtan and Beijing Changxindian. *China City Planning Review, 2008*(4), 45–53.

CHAPTER 7

# Conclusion and Discussion

## 7.1 Summary of Main Findings and Arguments

The first half of this research has responded two main research questions—eco-city ought to be seen as the humanistic planning, and how to introduce more humanistic planning and design principles into Chinese eco-city planning. The second half of this research focuses on the implication of people–place interactive relationship, that is, in what way and to what extent has this kind of relationship, known as the behavioural connection between people and place, been seen as a possible humanistic approach to realizing the people-oriented planning goal.

In Chapter 2, it discusses the people-oriented concept by reviewing different theories in three disciplines—urban planning, urban design and applied social psychology. It integrates these three theories into the integrated theoretical framework to better illustrate the humanistic planning under the context of ecology. The integrated theoretical framework itself improves the existing people-oriented planning concept in several ways. Firstly, it examines the concept of humanistic planning, specifically in urban planning and design area, and also broadly in terms of extending beyond the typical focus on psychological area. It suits *the people-oriented planning under the context of ecology* in the broader context with which it interacts, as the people-oriented eco-city planning goal is influenced by two different factors, namely people and environment. Secondly, this


Y. Wu, *Possibilities and Limitations of Chinese Eco-City Development*, New Perspectives on Chinese Politics and Society,
https://doi.org/10.1007/978-981-95-1958-3_7

theoretical framework more explicitly builds the complex and dynamic nature of humanistic planning, such as people-oriented concept and the interactive people–place relationship, which are two core elements which will be further discussed in the discussion chapters.

In Chapter 3, the concept of people's needs has been deemed as a relatively mental and cultural notion by different subjects, such as economics, sociology and philosophy. The well-known model for illustrating human needs is the one proposed by Abraham Maslow in his paper *A Theory of Human Motivation* (1943). He used the five-level pyramid to reveal people's needs not only include physiological needs but also include psychological ones. And he believes people's behaviours in certain moments have been affected by the most severe needs. *The New-Type Conceptual Model of Human-Centred Design* reveals the product design evolution, from functionality and usability to desirability, which has responded to the hierarchy of human needs. Also, this model argues that to enter the human-focused design phase, design ought to satisfy different range of human needs, even some subtle needs which users have not recognized or unexpressed. By introducing the new-type conceptual model, this chapter argues people's needs matter because it is directly related to the people-centred design principle, and it also argues people's behaviour is the manifestation of higher-level (invisible) needs.

By combining Chapters 2 and 3, this research constructs the analytical framework based upon the existing literatures to determine the definition of people-oriented concept of this research—humanistic planning in an interdisciplinary context focuses on the interaction between environment and people (e.g. people measure the spatial environment by their behaviours), and its main objective is to explore people's needs (including visible and invisible ones) when they stay in the spatial environment.

In Chapter 4, the first section aims to give an overview of emerging eco-city concept in China and related planning policies to contextualize planning of the SSTEC. Then, it delineates some general information about the SSTEC project at the beginning of this chapter, including its geographical and political conditions, its bilateral supervisory management framework, which are basic but unique characteristics can affect further data collection and analysis of this research by demonstrating the project positioning of the SSTEC. In addition, this chapter focuses on how people-oriented concept reflects in Chinese eco-city development. Specially, from the perspective of urban planning of the SSTEC, the indicator system has continuously shown the ongoing trend of people-oriented development pattern. And from the perspective of urban design of the SSTEC, the masterplan has been formulated to provide a clear blueprint to guide the SSTEC's overall urban development to achieve its KPIs, and to offer people a life-convenient living settlement under the general people-oriented design concept. In the scope of second half of this chapter, it plans to introduce elements that can constitute responsive public spaces of the SSTEC, including permeability and legibility as well as variety and robustness, to argue people-oriented concept has been rooted in the level of concrete urban design means.

In Chapter 5, by comparing research data of Zhongxin Avenue and children's park, people regard the safety issue as the primary consideration of their sense of activity experience, significantly influencing their psychological feeling in public spaces—unsafe environment makes people hold negative attitudes, such as uncomfortableness, towards spatial quality of this public space (e.g. Zhongxin Avenue), vice versa (e.g. children's park). That is, safety issue, mainly caused by the spatial layout of legibility and permeability, is regarded as the precondition to make people psychologically feel comfortable. Secondly, in Chapter 6, the variety of social infrastructure ties people and place together. Research data conducted in both the 3rd community centre and the 4th community centre shows, the direct cause of people's positive or negative feeling is precisely due to the condition of social infrastructures. People feel comfortable or convenient if the indoor environment has well air-supplying condition, while people feel uncomfortable or inconvenient if the outdoor environment has inadequate seating options and barrier-free access. These are two forms of the behavioural connection between people and place, and also are two *leverage points* of people's fundamental needs, psychologically and physically, when staying in the public space within the eco-city.

In Chapter 6, this research identifies the obstacle for meeting people's dynamic needs when living in the eco-city—the pressure about flexible adjustments for design features, such as additional manpower or financial resources have to be invested for eco-city's redevelopment to meet people's changing needs during the post-operational management stage. Last but not least, taking constraints on promoting as the starting point, this research proposes the possible improved eco-city's design process, from the traditional four steps to six steps based upon predicting and integrating *people's cognitive and behavioural pattern*, in order to mitigate limiting impacts on the conceptual-practical gap accompanying the realization of humanistic planning in the Chinese eco-city project.

The physical environment has implications for people's lives, and the purpose of the environment is to afford opportunities for achieving the quality of life. However, urban planning may not relate specifically to the particular needs and expectations for each individual. People are different and have different needs that one cannot specify, identify or posit one model of the people-environment relationship. That is, people are active and not passive; they interpret and evaluate the environment in ways that they wish to. Therefore, the term *quality of life* has been used to indicate the 'subjective judgement by an individual as to the degree to which her or his needs in the various domains of life are met' (Churchman, 2002, p. 7). Based upon the levels of *Maslow's Hierarchy of Needs Model*, these domains include the degree of self-actualization, social relations, public services, environmental quality and so on (Churchman, 1993). From the perspective of the integration of disciplines, namely from the point of view of the field of environmental psychology, the goal of humanistic urban planning is to balance the relationship between people and place as well as to enable people to achieve high-level quality of life. Achieving this goal is not simple, as Churchman (2002) suggested, it requires 'an understanding of the way in which people perceive and think, and an understanding of the ways in which the socio-physical environment can be an asset or a hindrance to this goal' (p. 8).

## 7.2 Reflections on Research Questions

This research tends to be the normative slant—it aims to compare the practical level of Chinese eco-city project with the conceptual precepts of humanistic planning. In this research, the literature review part plans to make an argument why humanistic planning should be introduced in Chinese eco-city's project, and why it would be the theoretical basis for people–place interactive relationship to make the abstract people-oriented planning concept easy to understand and to collect research data. By reviewing this concept in existing literatures, a theoretical framework indicating the humanistic planning under the idea of ecology has been established. Specially, the theoretical framework itself, deriving from *Garden City*, *Ecological Urbanism* and *Environmental Psychology*, improves the existing people-oriented planning concept as it examines the concept of humanistic planning specifically in urban planning and design area, and also broadly in terms of extending beyond the typical focus on psychological area. It suits *the people-oriented planning under the context of ecology* in the broader context with which it interacts, as the people-oriented eco-city planning goal is influenced by two different factors—people and environment. Based upon this framework, this research examines the particular people–place relationship in the eco-city project from perspectives of both residents and planning system to respond to the actual situation about the existence of conceptual-practical gap of people–place relationship identified in Chapter 5. In Chapter 6, it examines the main constraint and planning initiatives on promoting the behavioural connection between people and place, and by discussing the existing traditional eco-city's design process, possible improved design process in terms of recommendation for practice has been put forward to demonstrate the implication of behavioural connection between people and place in the eco-city planning development.

Research findings have highlighted concrete meanings of the people-oriented concept into Chinese eco-city planning practice, including the core connotation (people–place relationship) and implications (possible improved design process) in a broader sense of humanistic planning under the context of ecology. An initial question has been addressed at the beginning of this research—do we see any evidence that humanistic planning originated in the mainstream global theory has happened in Chinese eco-city's practices? Key findings referring to each research question are shown as follows.

On the one hand, as discussed in Chapter 1, the case of global eco-city has not formed a unified concept; Chinese eco-city planning practice is still in the early stage of practical exploration. It is easy to be confused with other sustainable concepts, such as liveable city, ecological garden city or low-carbon city, which results in both status and formulation methods of eco-city planning not yet being established. This research defines the meaning of eco-city concept in the context of Chinese green urbanization, in order to provide the referencing value for Chinese government to formulate development goals and guiding principles in the future. Being different from development models of eco-cities in other countries, the Chinese eco-city development represented by the SSTEC project is the corresponding product to adapt to the change of *New-Type Urbanization* development pattern—the gradual transformation from the newly built city construction which mainly focuses on economic and environmental development under the traditional urbanization pattern to the trinity of economic-environmental-humanistic urban governance which is characterized by the concept of *humanistic planning under the context of ecology* under the green urbanization pattern. This research regards the construction of eco-city to be the driving force for the current stage of Chinese *New-Type Urbanization*—under the carbon peaking and carbon neutrality goal as well as the digital reform goal, for Chinese eco-city, the people-oriented planning concept is the basis of its development, low-carbon and digital technology are the way to its sustainable development, and creating the liveable environment for residents is its ultimate development goal.

On the other hand, the physical environment has implications for people's lives, and the purpose of physical environment is to afford opportunities for achieving the quality of life. However, urban planning may not relate specifically to the particular needs and expectations for every individual. People are different and have different needs that one cannot specify, identify or posit one model of the people-environment relationship. That is, people are active and not passive; they interpret and evaluate the environment in ways that they wish to. Therefore, the term *the quality of life* has been used to indicate the 'subjective judgement by an individual as to the degree to which her or his needs in the various domains of life are met' (Churchman, 2002, p. 7). Based upon the levels of *Maslow's Hierarchy of Needs Model*, these domains include the degree of self-actualization, social relations, public services, environmental quality and so on (Churchman, 1993). From the perspective of disciplinary integration, by introducing environmental psychology theory into urban planning field, it contributes to realize the goal of humanistic urban planning—to balance the relationship between people and place as well as to enable people to achieve high-level quality of life. Achieving this goal is not simple, as Churchman (2002) suggested, it requires 'an understanding of the way in which people perceive and think, and an understanding of the ways in which the socio-physical environment can be an asset or a hindrance to this goal' (p. 8).

## 7.3 Original Research Contributions

Findings of this research hold some implications for understanding people-oriented planning concept in the realm of eco-city development. Broadly construed, this research represents an original contribution to knowledge, manifesting at the theoretical level with some empirical dimensions. At the theoretical level, it establishes connections between psychological theories and theories pertaining to urban planning and design. Specially, this research extends dialogues surrounding people-oriented planning by delving into people's psychological needs. At the empirical level, it introduces the analysis of mainstream humanistic planning concept by conceptualizing the people–place relationship. Additionally, this research presents insights garnered from the SSTEC initiative, an area that has received limited attention referring to the social sustainability aspect of eco-city development in academic discourse. As an interdisciplinary study, it enriches some research domains, including eco-city literatures, the combined analytical framework and practical contributions about people-oriented planning. Detailed illustration of these three aspects has been shown as follows.

### *7.3.1 Eco-City Literatures*

This research contributes to a body of critical literature about people-oriented eco-city initiatives, which aims to seek for identifying the conceptual-practical gap accompanying the realization of humanistic planning under the idea of ecology. The term *eco-city*, in this research, has been seen as the *symbolic* to describe a broader range of Chinese contemporary urban-scale initiatives, which attempts to pursue the planning goal of sustainable and humanistic development pattern. Such initiatives are worthy of study because of the increasing international mainstream into eco-city's planning goal and its policy-making process since the turn of the century (Joss, 2011; Joss et al., 2013). As the global body of discourse and practice with enough support of considerable financial resources and political wills, the eco-city has represented a potent collective perception about social sustainability, and the hope for planning better living environment for people.

Since the boom of eco-city in recent years, its critiques from a variety of angles have been raised by the academia. For instance, the disconnection between eco-city development and wider urban context (Caprotti,

2014; Caprotti et al., 2015; Hodson & Marvin, 2010; Joss & Molella, 2013), the feasibility about the 'borrowed' global eco-city plans into local contexts (Chang & Sheppard, 2013; Hult, 2013; Rapoport, 2015), the governance processes through which eco-city initiatives are implemented (Joss, 2011; Joss et al., 2015), as well as the potential conflicts between technological showcases and sustainability in the eco-city development (Carvalho, 2015; Shwayri, 2013; Yigitcanlar & Lee, 2014). This research regards the last critical perspective—technological overuse as the starting point to identify potential limitations of the contemporary Chinese eco-city development, but its original contribution lies in a shift of principle emphasis to the nature of *people-oriented eco-city* as envisaged in, and resulting from related broader goal of sustainable development by the case study of Chinese pilot project. It argues that *people-oriented planning under the context of ecology* concept is more than an approach, or 'adding green' for enabling people to enjoy the nature. As the case study reveals in this research, while public spaces of the SSTEC have provided ecologically and socially meaningful places for living residents, high green coverage focusing on the fixed technical standards driven by the KPIs still cannot completely reach its planning goal about meeting people's ever-changing and dynamic needs, especially their psychological feelings when staying in the open space. Eco-city, as one of the humanistic planning types, is a way of thinking about in what ways design issues limit people's needs to be met when staying in the built environment within the eco-city. Moreover, it is also a way of framing solutions to discover *leverage points* of the built environment to achieve the goal of harmonious people–place interactive relationship.

### *7.3.2 The Combined Analytical Framework*

Regarding the theoretical implication, this study combines both humanistic and ecological elements from different disciplines, to build the fundamental conceptual and analytical framework for guiding the whole research process. Such framework contributes to broadening the existing understanding about people-oriented planning concept—to highlight people's needs about seeking for better living environment not only include the physical aspect such as social infrastructures, but also require the psychological aspect such as positive feelings when people stay in the built environment. Moreover, by recognizing the significance of people's perception and behaviours in the humanistic planning concept, this research more explicitly builds that principle into eco-city's design process—behavioural connection has been found to be linked with the relationship between people and place. The new design model which integrates *people's cognitive and behavioural pattern* into other place-making procedures, can provide a possibility about analysing the rational derivative process that occurs in the mind of designers in order to clarify and to ensure different eco-city's design processes all put people's needs and behavioural preferences at the heart of planning and design work.

### 7.3.3 *Practical Contribution About People-Oriented Planning*

In spite of analysing eco-city project by using the appropriate humanistic concept, it is still difficult to find the optimal solution for (newly built) eco-city practices. There will still be a number of eco-city projects with similar issues discovered in the case study. By researching the SSTEC pilot project, specific constraints on effectively promoting the people-oriented concept have been clarified. For realizing people-oriented planning goal under the complex process of urban development, it is vital to have the effective multi-perspective planning and design approach—not only have a multi-disciplinary planning team, but also involve scholars and experts in different fields to participate in the design process or to establish an interdisciplinary communication mechanism. Such interdisciplinary collaborative approach takes the systemic and holistic interconnectedness between people and place into consideration, not only be useful in the eco-city development, but also can be applied to other types of Chinese cities, as people-oriented concept has become the mainstream trend of thought in Chinese urban development. From this perspective, humanistic planning under the idea of ecology, put forward in this research, contributes to the practical advocate from an interdisciplinary perspective within the study of eco-city—urban planning and design method is a holistic and all-embracing approach, rather than a simple 'adding' of separate planning ideas only by planning practitioners.

## 7.4 Limitations of the Research and the Need for Future Research

By establishing the theoretical framework *humanistic planning under the context of ecology* to examine the SSTEC case study, this research has provided insightful explanations concerning the practical meaning of humanistic planning from case-based evidences, such as the relationship between people and place that matters to implement people-oriented planning concept in Chinese newly built eco-city project. Also, it has shown the *conceptual-practical gap* to argue whether there are any evidences to show the global mainstream people-oriented planning concept has been implemented in the Chinese eco-city development. Future research might find this research helpful to conduct a further case study, which adopts by the interdisciplinary research perspective, to examine the occurrence and operation of people-oriented planning concept in other newly built eco-city projects from the lens of living experiences of residents.

Nevertheless, the research-based knowledge produced from findings also has limitations when applied into the real-world eco-city practical projects. This research only focuses on the applicability of mainstream humanistic theories to Chinese planning practices, but does not concentrate on the output or outcome of relevant projects. Such limitation is mainly because of the original aim of doing this research, and is also restricted by the time limit of the case study. Moreover, even though the people-oriented planning concept has already been proposed in Chinese national and local official documents as one of the main development goals, in the practical eco-city project, residents still have a variety of dissatisfaction about the spatial design, and the disharmony in the relationship between people and place still exists. Therefore, some of the research findings cannot be fully generalized to other eco-city cases as the main aim of this research is to put forward an advocacy conceptual interpretation of people-oriented planning theory, which has paid much attention to the relationship between people and place, and has argued eco-city planning should not only care about people's invisible needs but also people's invisible needs reflected by their behaviours. And it is of great importance for further research conducted in the future to comprehensively explore constraints on effectively promoting the people-oriented concept in Chinese eco-city project, taking the procedural limitation of Chinese urban planning and design mechanism as the starting point. Moreover, the author worked for *Zhejiang Development and Planning Institute* in the year of 2022 and participated in the urbanization construction project of Xinchang County (新昌县). This place aims to build an exquisite garden city (打造精致花园城市); the author plans to combine the content of this book with future research about the development of Xinchang County.

Although this case study examined the conceptual-practical gap of the SSTEC project since 2019, this six-year research can only represent planning and design issues in recent years since changes and dynamics of people's needs will then be exposed and evolve further over time. Thus, plenty of scopes for future's more extensive examinations about the changing trend of humanistic planning in the eco-city development, and also about introducing more psychological knowledge into the discipline of urban planning and design need to be researched. Moreover, as is common in qualitative research, some limitations in the use of the methodology are inevitable. For example, in terms of the questionnaire survey, more responses secured, more accurate results can be received (Bryman, 2016). However, the respondent (residents in the eco-city) is a randomly selected population in the limited research time; consequently, it has been difficult to get responses from a significantly larger number of street-intercept questionnaires. Similar situation also applies to the in-depth interviews.

## 7.5 Conclusion

Firstly, this chapter has summarized main research findings, including the empirical evidence that derived from constructing the analysis framework to the case study, and also the explanation about how research questions have been answered. Secondly, it elaborates theoretical, methodological and practical contributions of this research, from the perspective of interdisciplinary research. Last but not least, this chapter indicates some limitations in terms of its generalizability and research methods, which helps to create more scopes for future studies.

## References

Bryman, A. (2016). *Social research methods* (5th ed.). Oxford University Press.

Caprotti, F. (2014). Critical research on eco-cities? A walk through the Sino-Singapore Tianjin eco-city, China. *Cities, 36*, 10–17.

Caprotti, F., Springer, C., & Harmer, N. (2015). 'Eco' for whom? Envisioning eco-urbanism in the Sino-Singapore Tianjin eco-city, China. *International Journal of Urban and Regional Research, 39*(3), 495–517.

Carvalho, L. (2015). Smart cities from scratch? A socio-technical perspective. *Cambridge Journal of Regions, Economy and Society., 8*(1), 43–60.

Chang, I.-C.C., & Sheppard, E. (2013). China's eco-cities as variegated urban sustainability: Dongtan eco-city and Chongming eco-island. *Journal of Urban Technology., 20*(1), 57–75.

Churchman, A. (1993). *A differentiated perspective on urban quality of life: Women, children and elderly.* Cited in: Bonnes, M. (1993). *Perception and evaluation of environmental quality* (pp. 165–178), 1st edn. Rome: UNESCO Programme on Man and Biosphere.

Churchman, A. (2002). Environmental Psychology and Urban Planning: Where Can the Twain Meet. https://www.researchgate.net/publication/284027102 Accessed on 10th December 2022.

Hodson, M., & Marvin, S. (2010). Urbanism in the anthropocene: Ecological urbanism or premium ecological enclaves? *Cities, 14*(3), 298–313.

Hult, A. (2013). Swedish production of sustainable urban imaginaries in China. *Journal of Urban Technology., 20*(1), 77–94.

Joss, S. (2011). Eco-cities: The mainstreaming of urban sustainability: Key characteristics and driving factors. *International Journal of Sustainable Development Planning., 6*(2), 1–18.

Joss, S., Cowley, R., & Tomozeiu, D. (2013). Towards the 'Ubiquitous Eco-City': An analysis of the internationalisation of eco-city policy and practice. *Urban Research and Practice, 6*(1), 54–74.

Joss, S., & Molella, A. P. (2013). The eco-city as urban technology: Perspective on Caofeidian international eco-city, China. *Journal of Urban Technology, 20*(1), 115–137.

Joss, S., Cowley, R., de Jong, M., Müller, B., Park, B.-S., Rees, W., Roseland, M. & Rydin, Y. (2015). Tomorrow's City Today: *Prospects for Standardising Sustainable Urban Development*. 1st ed. London: University of Westminster

Maslow, A. H. (1943). A theory of human motivation. *Psychological Review, 50*(4), 370–396.

Rapoport, E. (2015). Globalising sustainable urbanism: The role of international master planners. *Area, 47*(2), 110–115.

Shwayri, S. (2013). A model Korean ubiquitous eco-city? The politics of making Songdo. *Journal of Urban Technology, 20*(1), 39–55.

Yigitcanlar, T., & Lee, S.-H. (2014). Korean ubiquitous-eco-city: A smart-sustainable urban form or a branding hoax? *Technological Forecasting & Social Change, 89*, 100–114.

# Appendix A: Fieldwork Proposal

**【Participant Observation】**

1. **Research Object**

Residents and physical environment of eco-city.

2. **Concerns**

   (1) Types of outdoor activity in public space
   (2) Users of public space (e.g. different ages and occupations)
   (3) How people use the public space (their using time and using frequency)
   (4) People's attitudes towards the spatial quality of public space

3. **Research Process**

   (1) Public space analysis: this research work is mainly done by visiting and observing chosen public spaces within the eco-city, and it focuses on the following questions:
      – Which public spaces within the eco-city can be used for doing outdoor activities?
      – Where do residents need public spaces in the eco-city?
      – What qualities do the existing public spaces have in the eco-city?



Y. Wu, *Possibilities and Limitations of Chinese Eco-City Development*, New Perspectives on Chinese Politics and Society,
https://doi.org/10.1007/978-981-95-1958-3

– How to explore the potential qualities and problems of public spaces have in the eco-city?

(2) Public life survey: it focuses on studying how public spaces within the eco-city are used, and what people's activities are carried out in these spaces. This research work is conducted at specific times and seasons (summer and winter, both working days and weekends, and weather suitable for travel), by observing and recording residents' outdoor life and activities within the chosen places of the eco-city. Moreover, in order to obtain detailed and in-depth understanding of residents' attitudes towards satisfaction of public spaces in eco-city and their needs of social facilities, questionnaires are randomly distributed within the chosen places of the eco-city.

4. **Early-Stage Preparation**

(1) Items: stopwatch, counter, pen, clipboard, notebook, camera, etc.

(2) Material: risk assessment form, ethical approval form, detailed location map of the research area (*Sino-Singapore Tianjin Eco-City*), site map for marking (including road boundary line, street name, building entrance, seats and other social facilities) and pre-made recording template (including the number of people as well as their gender and age, detailed information about their activities).

(3) Seasons and time selection: in order to ensure the objectivity of research data, seasons are selected for good weather in winter and summer, and the time covers both working days and weekends (at least two each), usually from morning to midnight (i.e. 8 am–6 pm, 6 pm–12 am). One hour as a time unit, and each unit will be recorded and counted for 15 minutes.

5. **Specific Methods of Investigation**

Specific operational approaches of participant observation include mapping, (on-site) counting, photographing, diary keeping, street-intercept survey (questionnaire) and in-depth interviews. Each method has its own emphasis and complements other methods, which contribute to obtaining the comprehensive research data about public space and public life in the eco-city.

(1) **Mapping**

- To accurately reflect the status of public life, this method plans to use different symbols to record a variety of people's activities within the selected public spaces, by means of observing and marking those activities on the pre-prepared site map.
- **Site selection:** observation aims to select not only the crowded areas, including spaces with stationary activities (e.g. sitting and standing) and spaces with active activities (e.g. exercising and playing), but also the neglected areas, such as plaza space and street space within the eco-city.
- **Data collection:** time period covered by data collection of mapping is similar to the on-site counting method. It aims to record different types of people's activities in the selected areas of eco-city, and also the number of people who participate in those activities. As this approach focuses on understanding the status of public life, passing pedestrians in the selected areas are not counted, that is, the main research object of both mapping and counting is the residents of eco-city. Specifically, it concerns different types of people's activities, including stopping and staying, waiting for the transport, as well as those participating in the commercial, cultural, sports activities. It plans to use the pre-prepared site map and temple to record data.
- **Data analysis:** at this stage, it plans to import the recorded research data into SPSS, then to use this software to do the preliminary analysis. Data analysis mainly involves the following aspects:
  - Types and the number of activities during working days (day and night, winter and summer) in one specific public space
  - Types and the number of activities during weekends (day and night, winter and summer) in one specific public space.
  - Comparative analysis about the number of activities during working days and weekends (day and night, winter and summer) in one specific public space.

- Comparative analysis about the number of activities in different public spaces per 15 minutes.
- Comparative analysis about the number of activities at the given time period (e.g. rush hour, lunch time) in one specific public space.
- Comparative analysis about the average number of activities in winter and summer in one specific public space.

- **Data application:** final results of the research data about mapping are used to describe the current situation of different public spaces within eco-city, that is, by whom they are used, which are the most frequent activities taken place within eco-city, and also the average lasting time of each outdoor activity.

(2) **On-Site Counting**

- On-site counting focuses on understanding conditions of both pedestrian and walking environment. By recording the number, gender and age (0–6 years old, 7–14 years old, 15–24 years old, 25–44 years old, 45–64 years old and more than 65 years old) of walkers, this research method aims to objectively reflect how walkers use public spaces within the eco-city.
- **Site selection:** observation aims to select major streets, urban squares, community parks, where can almost reflect conditions and objective laws of people's outdoor activities. Site selection is randomly targeted within the central areas of eco-city. And in order to facilitate the data collection, research areas are chosen near the three community centres, where the eco-city is densely populated.
- **Data collection:** time period covered by data collection of on-site counting is similar to the mapping method. It aims to record the pedestrian flow in the selected areas of eco-city (day and night, winter and summer), including the number of pedestrians per minute and the number of pedestrians per 15 mins. While recording, it needs to indicate the age and gender of pedestrians on the pre-made recording template.

- **Data analysis** at this stage plans to import the recorded research data into SPSS, then to use this software to form some comparative charts for preliminary analysis. And data analysis mainly involves the following aspects:
  - The average number of pedestrians at different times in one hour in one specific area.
  - Comparative analysis about the number of pedestrians between daytime of working days and daytime of weekends in one specific area.
  - Comparative analysis about the number of pedestrians between night of working days and night of weekends in one specific area.
  - Comparative analysis about the number of pedestrians between daytime of working days and night of working days in one specific area.
  - Comparative analysis about the number of pedestrians between daytime of weekends and night of weekends in one specific area.
- **Data application**: final results of the research data about traffic flows are not only used to describe the grade status of roads, but also to reflect the quality of walking environment, namely the spatial quality. That is, even though there are lots of people walking on the street, it does not mean this walking environment has a high spatial quality. Only when most of people prefer to actively stay on the pedestrian streets and enjoying public life there, can indicate the high spatial quality of this place. And those comparative analysis of research data can reflect the use of pedestrian streets, as well as the basic laws of flow changes.

(3) **On-Site Observation**

- By conducting research field work, the main purpose of on-site observation is to investigate the physical environment for getting information about what spatial elements can stimulate people to spend more time in public spaces. Observation is based upon 12 quality criteria proposed by Jan Gehl team, and they are structured around three main themes, namely protection, comfort and enjoyment. Firstly, without basic protection from cars, noise, rain and wind, people will not be willing to spend more time in a space. Secondly, without elements that make walking, standing, sitting, seeing and conversing comfortable, a place generally will not invite more people to spend time there. Thirdly, the pleasing public space tends to offer positive aesthetic and sensory experiences, to take advantage of local climate condition (e.g. offering shade in hot summer), and to provide human-scale elements so that people will not feel lost in their surroundings.

- **Site selection:** selected places, including urban plazas, community parks and streets, should be covered, same as the sites of mapping and counting. Public spaces with low activity are not included to assess.
- **Data collection:** this research investigation aims to be done at an active time of day with good weather. However, it could also occur at any time of the day, on any day of the week, since it looks at spatial elements that do not change very much. When getting to the site, take about 5 minutes to simply observe for better knowing about the place. Start where most people spend time within this place. Assessing and taking notes of what has been seen, then giving score (3 means yes, 2 means in between, 1 means no) for each criterion.
- **Data application**: final results of the research data are presented in the summary of observational diary, being used for an objective description about the physical environment of eco-city. Moreover, research data from on-site observation is also used as a reference to formulating questionnaire and in-depth interview questions for both residents and planning practitioners.

**【Street-Intercept Survey (Qualitative Questionnaire)】**

1. **Research Object:** residents of the eco-city.
2. **Concerns**

(1) Why do people come here? What activities will they do?

(2) Regarding the comfortable and uncomfortable of public spaces/ streets, what opinions do they hold? And what are suggestions for improving the comfort of public spaces in the future?
(3) What are factors that affect the vitality of public spaces?
(4) The frequency of people coming to these public spaces, and how long are they staying there.
(5) People's like and dislike aspects of spatial elements.
(6) People's sense of security about these public spaces.

3. **Data Collection and Application**
Site selection is targeted within the central areas of eco-city. And in order to facilitate the data collection, research areas are chosen near the three community centres, where the eco-city is densely populated. Semi-structured questionnaires are randomly distributed to residents during participant observation within the selected public spaces. Final results of the research data are presented in the summary of writing report.

**【In-Depth Interviews】**

1. **Research Object**
Planning practitioners of *Sino-Singapore Tianjin Eco-City* and project operators.
2. **Concerns**
(1) How revised indicator system reflect the people-oriented concept.
(2) The embodiment of people-oriented concept in SSTEC project.
3. **Data Analysis**
It aims to use the grounded theory for data coding. And the summary of *themes* is used for extracting interviewing information, and the logical thinking based on research questions is applied to recombine relationship among those *themes*.

# Appendix B: The Template of People's Numbers and Ages

**Location: Date:**
**Time Slot(15 minutes):**
**Male:**

| *Age group* | *People counting* | *Total number* |
|---|---|---|
| 0–6 | | |
| 7–14 | | |
| 15–24 | | |
| 25–44 | | |
| 45–64 | | |
| Over 65 | | |

**Female**:

| *Age group* | *People counting* | *Total number* |
|---|---|---|
| 0–6 | | |
| 7–14 | | |
| 15–24 | | |

(continued)



Y. Wu, *Possibilities and Limitations of Chinese Eco-City Development*, New Perspectives on Chinese Politics and Society,
https://doi.org/10.1007/978-981-95-1958-3

(continued)

| *Age group* | *People counting* | *Total number* |
|---|---|---|
| 25–44 | | |
| 45–64 | | |
| Over 65 | | |

# Appendix C: City at Eye Level: 12 Quality Criteria



Y. Wu, *Possibilities and Limitations of Chinese Eco-City Development*, New Perspectives on Chinese Politics and Society,
https://doi.org/10.1007/978-981-95-1958-3

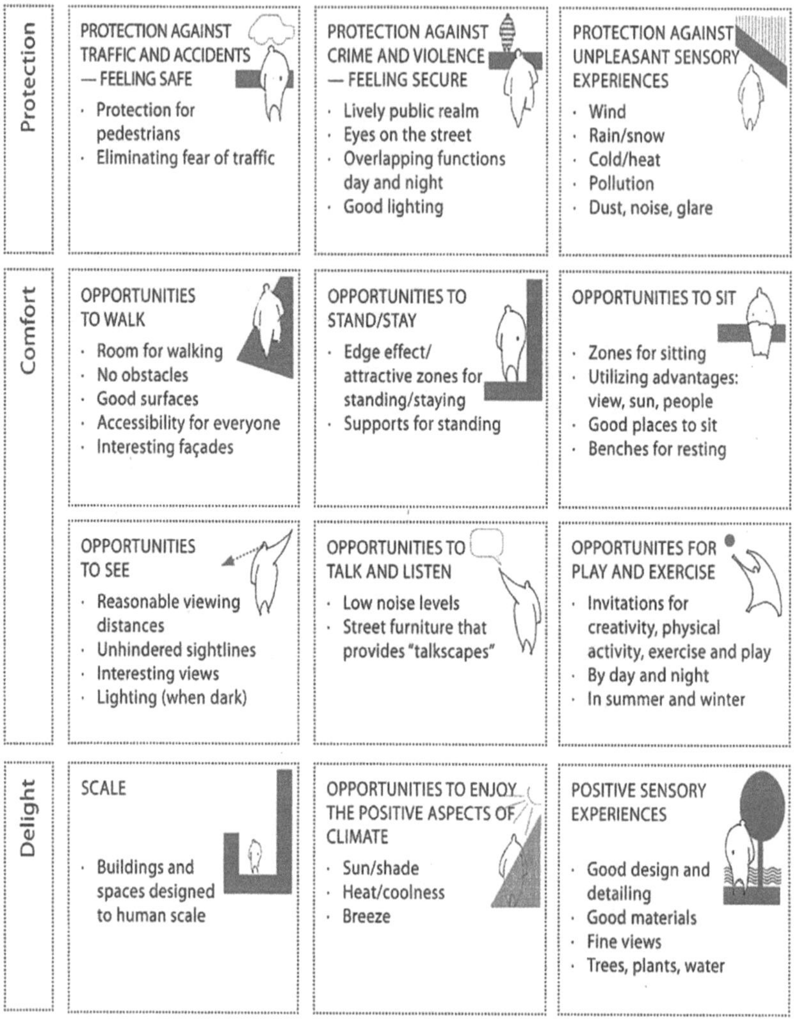


(Source: Figure is from Gehl Architects—Urban Quality Consultants, 2009.)

# Appendix D: Specific Questions Generated by 12 Quality



Y. Wu, *Possibilities and Limitations of Chinese Eco-City Development*, New Perspectives on Chinese Politics and Society,
https://doi.org/10.1007/978-981-95-1958-3

| *Location: 3 = Yes, 2 = In between, 1 = No* | | | |
|---|---|---|---|
| ***Protection*** | *Protection against traffic and accidents*<br>Do groups across age and ability experience traffic safety in the public space?<br>Can people safely bike and walk without fear of being hit by a driver? | *Protection against harm by others*<br>Is the public space perceived to be sage both day and night?<br>Are there people and activities at all hours of the day because the area has, for example, both residents and offices?<br>Does the lighting provide safety at night? | *Protection against unpleasant experience*<br>Are there noises, smells, or other pollution?<br>Does the public space function well when it is windy?<br>Is there shelter from strong sun and rain? |
| ***Comfort*** | *Options for mobility*<br>Is this space legible and accessible?<br>Are there any physical elements that can impact personal mobility in the forms of walking, using of a wheelchair, or pushing a stroller? | *Options for standing*<br>Does the place have spatial features that people can stay and lean on, like a facade that invites one to spend time next to it, a bus stop, a bench, a tree, or a small ledge or niche? | *Options for sitting*<br>Are there good primary seating options, such as benches or chairs?<br>Or is there only secondary seating, such as a stair, seat wall, or the edge of a fountain? |
| | *Options for seeing*<br>Does the place have interesting things for people to look at? | *Options for talking and listening/hearing*<br>Do people have options to sit together for a conversation? | *Options for play, exercise and activities*<br>Are there options to be active at multiple time of the day and year? |
| ***Enjoyment*** | *Scale*<br>Is the public space and the building that surrounds it at human scale? | *Opportunities to enjoy the positive aspects of climate*<br>Are local climatic aspects, such as wind and sun taken into account? | |

(Source: Information is from Gehl Architects—Urban Quality Consultants, 2009.)

# Appendix E: Summary of Descriptions About Data

## Classification of People's Activities in Each Research Site

### People's Activities in Zhongxin Avenue

| *People's activity* | *Types of activity* |
|---|---|
| Walking between bus stop and destination | Necessary activity |
| Waiting for bus | Necessary activity |
| Sitting and resting | Optional activity |

It is obvious the majority of people's activities in Zhongxin Avenue is necessary activity.

### People's Activities in the 3rd Community Centre

| *People's activity* | *Types of activity* |
|---|---|
| Standing and chatting | Optional activity and social activity |
| Waiting for bus | Necessary activity |
| Sitting and resting | Optional activity |
| Business activities | Social activity |
| Baby sitting | Optional activity and social activity |
| Children playing | |



Y. Wu, *Possibilities and Limitations of Chinese Eco-City Development*, New Perspectives on Chinese Politics and Society,
https://doi.org/10.1007/978-981-95-1958-3

It is obvious that people's activities in the 3rd community centre are diverse, but the majority are optional and social ones.

**People's Activities in the Eco-Valley**

| *People's activity* | *Types of activity* |
|---|---|
| Standing and chatting | Optional activity and social activity |
| Sitting on the bench | Optional activity |
| Children playing | Social activity |
| Walking and doing exercise | Optional activity |

It is obvious the majority of people's activities in the eco-valley (2nd community centre) are optional and social ones.

**People's Activities in the Children's Park**

| *People's activity* | *Types of activity* |
|---|---|
| Standing and chatting | Optional activity and social activity |
| Sitting in chairs | Optional activity |
| Children playing | Social activity |
| Doing exercise (Tai Ji) | Optional activity |

It is obvious the majority of people's activities in the children's park are optional and social ones.

## Research Data Analysis (Zhongxin Avenue)

### *Summary of Public Life Research Data*

**Recording of People's Activities in Zhongxin Avenue**

***Date:*** *Thursday, 20 December 2018*
***Time:*** *8am–10am*
***Weather:*** *Sunny*

| *Record symbol* | *Types of activities* | *Number of people* | *Average staying time* | *Major age group* |
|---|---|---|---|---|
| • | Walking between bus stop and destination | 315 | 10 mins | The retired elderly, young people |
| T | Waiting for bus | 209 | 15 mins | The retired elderly, young people |
| □ | Sitting and resting (secondary seating, street stairs) | 13 | 15 mins | The Retired Elderly |

**Total Number of Recording People:**
315 + 209 + 13 = 537

***Date:*** *Thursday, 20 December 2018*
***Time:*** *1 pm–3 pm*
***Weather:*** *Sunny*

| *Record symbol* | *Types of activities* | *Number of people* | *Average staying time* | *Major age group* |
|---|---|---|---|---|
| • | Walking between bus stop and destination | 113 | 10 mins | The retired elderly |
| T | Waiting for bus | 93 | 20 mins | The retired elderly |
| □ | Sitting and resting (secondary seating, street stairs) | 17 | 20 mins | The retired elderly |

**Total Number of Recording People:**
113 + 93 + 17 = 223

***Date:*** *Thursday, 20 December 2018*
***Time:*** *6 pm–8 pm*
***Weather:*** *Sunny*

| *Record symbol* | *Types of activities* | *Number of people* | *Average staying time* | *Major age group* |
|---|---|---|---|---|
| • | Walking between bus stop and destination | 483 | 5 mins | Young people |
| T | Waiting for bus | 279 | 15 mins | The retired elderly, young people |

**Total Number of Recording People:**
483 + 279 = 762

***Date:*** *Saturday, 22 December 2018*
***Time:*** *8am–10am*
***Weather:*** *Sunny*

| *Record symbol* | *Types of activities* | *Number of people* | *Average staying time* | *Major age group* |
|---|---|---|---|---|
| • | Walking between bus stop and destination | 366 | 10 mins | The retired elderly |
| T | Waiting for Bus | 241 | 15 mins | The retired elderly |
| □ | Sitting and resting (secondary seating, street stairs) | 26 | 15 mins | The retired elderly |

**Total Number of Recording People:**
366 + 241 + 26 = 633

***Date:*** *Saturday, 22 December 2018*
***Time:*** *1 pm–3 pm*
***Weather:*** *Sunny*

| *Record symbol* | *Types of activities* | *Number of people* | *Average staying time* | *Major age group* |
|---|---|---|---|---|
| • | Walking between bus stop and destination | 416 | 10 mins | Young people |
| T | Waiting for bus | 322 | 20 mins | Young people |

**Total Number of Recording People:**
416 + 322 = 738

***Date:*** *Saturday, 22 December 2018*
***Time:*** *6 pm–8 pm*
***Weather:*** *Sunny*

| *Record symbol* | *Types of activities* | *Number of people* | *Average staying time* | *Major age group* |
|---|---|---|---|---|
| • | Walking between bus stop and destination | 257 | 5 mins | Young people |
| T | Waiting for bus | 132 | 10 mins | Young people |

**Total Number of Recording People:**
257 + 132 = 389

## *The Summary of High-Frequency Adjective Words*

| *Adjective word ranking* | *Frequency* | *Evidences* |
|---|---|---|
| Unsafe | 75 | Narrow sidewalk<br>Lack of barrier between motorway and sidewalk |
| Uncomfortable | 62 | Lack of resting benches in the bus stop. The elderly people and pregnant women have to stand for a long |

(continued)

(continued)

| *Adjective word ranking* | *Frequency* | *Evidences* |
|---|---|---|
| Desolate | 57 | In winter, ornamental plants are untreated, flowers withered |

## Research Data Analysis (The 3rd Community Centre)

### *Summary of Public Life Research Data*

### People's Outdoor Activities in 3rd Community Centre

***Date:*** *Wednesday, 17 July 2019*
***Time:*** *8am–10am*
***Weather:*** *Sunny*

| *Record symbol* | *Types of activities* | *Number of people* | *Average staying time* | *Major age group* |
|---|---|---|---|---|
| • | Standing and chatting | 103 | 15 mins | The retired elderly |
| T | Waiting for bus | 96 | 15 mins | The retired elderly |
| □ | Sitting and resting (secondary seating, stair and the edge of landscape border) | 45 | 30 mins | The retired elderly |
| ○ | Business activities (distributing advertisements) | 24 | 90 mins | Young people |
| △ | Baby sitting and children playing | 18 | 30 mins | Children |
| **Total Number of Recording People:** 103 + 96 + 45 + 24 + 18 = 286 | | | | |

***Date:*** *Wednesday, 17 July 2019*
***Time:*** *1 pm–3 pm*
***Weather:*** *Sunny*

| *Record symbol* | *Types of activities* | *Number of people* | *Average staying time* | *Major age group* |
|---|---|---|---|---|
| • | Standing and chatting | 126 | 20 mins | The retired elderly |
| T | Waiting for bus | 73 | 15 mins | The retired elderly |
| □ | Sitting and resting (secondary seating, stair and the edge of landscape border) | 41 | 20 mins | Young people |
| ○ | Business activities (distributing advertisements) | 31 | 60 mins | Young people |

**Total Number of Recording People:**
126 + 73 + 41 + 31 = 271

***Date:*** *Wednesday, 17 July 2019*
***Time:*** *6 pm–8 pm*
***Weather:*** *Sunny*

| *Record symbol* | *Types of activities* | *Number of people* | *Average staying time* | *Major age group* |
|---|---|---|---|---|
| • | Standing and chatting | 91 | 20 mins | Middle-aged women |
| T | Waiting for bus | 62 | 10 mins | Young people |
| □ | Sitting and resting (secondary seating, stair and the edge of landscape border) | 47 | 30 mins | Middle-aged women |
| ○ | Business activities (distributing advertisements) | 18 | 45 mins | Young people |
| △ | Baby sitting and children playing | 24 | 30 mins | Children |

**Total Number of Recording People:**
91 + 62 + 47 + 18 + 24 = 242

***Date:*** *Saturday, 20 July 2019*
***Time:*** *8am–10am* ***Weather:*** *Cloudy*

| *Record symbol* | *Types of activities* | *Number of people* | *Average staying time* | *Major age group* |
|---|---|---|---|---|
| • | Standing and chatting | 58 | 20 mins | The retired elderly |
| T | Waiting for bus | 176 | 15 mins | Young people |
| □ | Sitting and resting (secondary seating, stair and the edge of landscape border) | 61 | 15 mins | The retired elderly |
| ○ | Business activities (distributing advertisements) | 35 | 90 mins | Young people |
| △ | Baby sitting and children playing | 29 | 30 mins | Children |

**Total Number of Recording People:**
58 + 176 + 61 + 35 + 29 = 359

***Date:*** *Saturday, 20 July 2019*
***Time:*** *1 pm–3 pm*
***Weather:*** *Cloudy*

| *Record symbol* | *Types of activities* | *Number of people* | *Average staying time* | *Major age group* |
|---|---|---|---|---|
| • | Standing and chatting | 81 | 30 mins | The retired elderly |
| T | Waiting for bus | 102 | 20 mins | Young people |
| □ | Sitting and resting (secondary seating, stair and the edge of landscape border) | 32 | 15 mins | The retired elderly |
| ○ | Business activities (distributing advertisements) | 26 | 90 mins | Young people |

**Total Number of Recording People:**
81 + 102 + 32 + 26 = 241

*Date: Saturday, 20 July 2019*
*Time: 6 pm–8 pm*
*Weather: Cloudy*

| *Record symbol* | *Types of activities* | *Number of people* | *Average staying time* | *Major age group* |
|---|---|---|---|---|
| • | Standing and chatting | 22 | 20 mins | The retired elderly |
| T | Waiting for bus | 48 | 15 mins | The retired elderly |
| □ | Sitting and resting (secondary seating, stair and the edge of landscape border) | 12 | 20 mins | The Retired Elderly |
| ○ | Business activities (distributing advertisements) | 13 | 60 mins | Young people |
| △ | Baby sitting and children playing | 8 | 20 mins | Children |
| **Total Number of Recording People:** 22 + 48 + 12 + 13 + 8 = 103 | | | | |

## *The Summary of High-Frequency Adjective Words*

| *Adjective word ranking* | *Frequency* | *Evidences* |
|---|---|---|
| Uncomfortable | 112 | No seating area for people to chat and to have a rest<br>Lack of shade, people feel too hot when they chat with others in the summer time<br>Among 112 copies, 73 copies have used hot (in summer) or cold (in winter) adjectives, due to the lack of indoor resting space, such as Starbucks |
| Inconvenient | 85 | Lack of entertainment and catering service for young people<br>Lack of barrier-free access at the entrance, being inconvenient for the elderly people and women with children |

# Research Data Analysis (Eco-Valley, Selected Community Section)

## *Summary of Public Life Research Data*

## Recording of People's Outdoor Activities in Eco-Valley

***Date:*** *Monday, 15 July 2019*
***Time:*** *8am–10am*
***Weather:*** *Sunny*

| *Record symbol* | *Types of activities* | *Number of people* | *Average staying time* | *Major age group* |
|---|---|---|---|---|
| • | Standing and chatting | 24 | 30 – 40 mins | Middle-aged women |
| □ | Sitting on the bench | 9 | 45 mins | Middle-aged women |
| ○ | Children Playing | 17 | 60 mins | Children |
| △ | Walking and doing exercise | 16 | 45 mins | The retired elderly |

**Total Number of Recording People:**
24 + 9 + 17 + 16 = 66

***Date:*** *Monday, 15 July 2019*
***Time:*** *1 pm–3 pm*
***Weather:*** *Sunny*

| *Record symbol* | *Types of activities* | *Number of people* | *Average staying time* | *Major age group* |
|---|---|---|---|---|
| • | Standing and chatting | 47 | 30 mins | The retired elderly |
| □ | Sitting on the bench | 23 | 45 mins | Middle-aged women |
| ○ | Children playing | 15 | 60 mins | Children |
| △ | Walking and doing exercise | 27 | 45 mins | The retired elderly |

**Total Number of Recording People:**
47 + 23 + 15 + 27 = 112

***Date:*** *Monday, 15 July 2019*
***Time:*** *6 pm–8 pm*
***Weather:*** *Sunny*

| *Record symbol* | *Types of activities* | *Number of people* | *Average staying time* | *Major age group* |
|---|---|---|---|---|
| • | Standing and chatting | 57 | 30–40 mins | Young people |
| □ | Sitting on the bench | 41 | 60 mins | Middle-aged women |
| ○ | Children playing | 36 | 45 mins | Children |
| △ | Walking and doing exercise | 62 | 45 mins | The retired elderly |

**Total Number of Recording People:**
57 + 41 + 36 + 62 = 196

***Date:*** *Saturday, 13 July 2019*
***Time:*** *8am–10am*
***Weather:*** *Sunny*

| *Record symbol* | *Types of activities* | *Number of people* | *Average staying time* | *Major age group* |
|---|---|---|---|---|
| • | Standing and chatting | 62 | 60 mins | Middle-aged women |
| □ | Sitting on the bench | 31 | 45 mins | Middle-aged women |
| ○ | Children playing | 57 | 60 mins | Children |
| △ | Walking and doing exercise | 86 | 30 mins | Young people |

**Total Number of Recording People:**
62 + 31 + 57 + 86 = 236

***Date:*** *Saturday, 13 July 2019*
***Time:*** *1 pm–3 pm*
***Weather:*** *Sunny*

| *Record symbol* | *Types of activities* | *Number of people* | *Average staying time* | *Major age group* |
|---|---|---|---|---|
| • | Standing and chatting | 14 | 20 mins | The retired elderly |

(continued)

(continued)

***Date:*** *Saturday, 13 July 2019*
***Time:*** *1 pm–3 pm*
***Weather:*** *Sunny*

| *Record symbol* | *Types of activities* | *Number of people* | *Average staying time* | *Major age group* |
|---|---|---|---|---|
| □ | Sitting on the bench | 19 | 45 mins | Middle-aged women |
| ○ | Children playing | 26 | 30 mins | Children |
| △ | Walking and doing exercise | 23 | 20 mins | The retired elderly |

**Total Number of Recording People:**
14 + 19 + 26 + 23 = 82

***Date:*** *Saturday, 13 July 2019*
***Time:*** *6 pm–8 pm*
***Weather:*** *Sunny*

| *Record symbol* | *Types of activities* | *Number of people* | *Average staying time* | *Major age group* |
|---|---|---|---|---|
| • | Standing and chatting | 28 | 30–40 mins | Young people |
| □ | Sitting on the bench | 33 | 45 mins | Middle-aged people |
| ○ | Children playing | 25 | 30 mins | Children |
| △ | Walking and doing exercise | 38 | 30–40 mins | Young people |

**Total Number of Recording People:**
28 + 33 + 25 + 38 = 124

## *The Summary of High-Frequency Adjective Words*

| *Adjective word ranking* | *Frequency* | *Evidences* |
|---|---|---|
| Safe | 117 | 1. The adult has reasonable viewing distance to keep eyes on their children<br>2. Activities of children's playing occur in the daytime, and adults' exercise occur in the night time, that is, space has different kinds of activities and users during day and night |
| Convenient | 109 | 1. The adult who take care of their children can also enjoy this public space; they sit on benches under the sunshade and can chat with neighbourhoods<br>2. Being regarded as the community park, this convenient location can benefit different user groups |
| Comfort | 102 | 1. Enough sitting space<br>2. Good fitness facilities, such as jogging track, plenty of trees |
| Happy | 93 | 1. The elderly can enjoy the sunshine in this public place and can chat with neighbours, especially for those retired people<br>2. Children can play with others and make friends<br>3. Young people can relax themselves after a long-day work |
| Open/Good views | 86 | 1. Open space, which has been suitable for human-scale activities<br>2. Adults can easily keep their eyes on their children playing nearby due to the reasonable viewing distance |

# Research Data Analysis (Children's Park)

## *Summary of Public Life Research Data*

### Recording of People's Outdoor Activities in Children's Park

***Date:*** *Tuesday, 16 July 2019*
***Time:*** *8am–10am*
***Weather:*** *Sunny*

| *Record symbol* | *Types of activities* | *Number of people* | *Average staying time* | *Major age group* |
|---|---|---|---|---|
| • | Standing and chatting | 8 | 20–30 mins | Middle-aged women |
| □ | Sitting in chairs | 28 | 60 mins | The retired elderly |
| ○ | Children playing | 30 | 60–90 mins | Children |
| △ | Doing exercise (Tai Ji) | 12 | 30–40 mins | The retired elderly |

**Total Number of Recording People:**
8 + 28 + 30 + 12 = 78

***Date:*** *Tuesday, 16 July 2019*
***Time:*** *1 pm–3 pm*
***Weather:*** *Sunny*

| *Record symbol* | *Types of activities* | *Number of people* | *Average staying time* | *Major age group* |
|---|---|---|---|---|
| • | Standing and chatting | 3 | 45 mins | Middle-aged women |
| □ | Sitting in chairs | 13 | 30 mins | Middle-aged women |
| ○ | Children playing | 15 | 30 mins | Children |

**Total Number of Recording People:**
3 + 13 + 15 = 31

***Date:*** *Tuesday, 16 July 2019*
***Time:*** *6 pm–8 pm*
***Weather:*** *Sunny*

| *Record symbol* | *Types of activities* | *Number of people* | *Average staying time* | *Major age group* |
|---|---|---|---|---|
| • | Standing and chatting | 35 | 60 mins | The retired elderly |
| □ | Sitting in chairs | 28 | 45 mins | The retired elderly |

(continued)

(continued)

***Date:*** *Tuesday, 16 July 2019*
***Time:*** *6 pm–8 pm*
***Weather:*** *Sunny*

| *Record symbol* | *Types of activities* | *Number of people* | *Average staying time* | *Major age group* |
|---|---|---|---|---|
| ○ | Children playing | 46 | 60 mins | Children |
| △ | Doing exercise (square dancing) | 56 | 60 mins | The retired elderly |

**Total Number of Recording People:**
35 + 28 + 46 + 56 = 165

***Date:*** *Sunday, 14 July 2019*
***Time:*** *8am–10am*
***Weather:*** *Cloudy*

| *Record symbol* | *Types of activities* | *Number of people* | *Average staying time* | *Major age group* |
|---|---|---|---|---|
| • | Standing and chatting | 10 | 45 mins | Middle-aged women |
| □ | Sitting in chairs | 19 | 60 mins | The retired elderly |
| ○ | Children playing | 25 | 60 mins | Children |

**Total Number of Recording People:**
10 + 19 + 25 = 54

***Date:*** *Sunday, 14 July 2019*
***Time:*** *1 pm–3 pm*
***Weather:*** *Cloudy*

| *Record symbol* | *Types of activities* | *Number of people* | *Average staying time* | *Major age group* |
|---|---|---|---|---|
| □ | Sitting in chairs | 8 | 40 mins | Middle-aged women |
| ○ | Children playing | 5 | 30 - 40 mins | Children |

**Total Number of Recording People:**
8 + 5 = 13

***Date:*** *Sunday, 14 July 2019*
***Time:*** *6 pm–8 pm*
***Weather:*** *Cloudy*

| *Record symbol* | *Types of activities* | *Number of people* | *Average staying time* | *Major age group* |
|---|---|---|---|---|
| • | Standing and chatting | 18 | 60 mins | The retired elderly |
| □ | Sitting in chairs | 28 | 45 mins | Middle-aged women |
| ○ | Children playing | 46 | 45 mins | Children |
| △ | Doing exercise (square dancing) | 76 | 60 mins | The retired elderly |

**Total Number of Recording People:**
18 + 28 + 46 + 76 = 168

## *The Summary of High-Frequency Adjective Words*

| *Adjective word ranking* | *Frequency* | *Evidences* |
|---|---|---|
| Safe | 82 | The adult has reasonable viewing distance to keep eyes on their children<br>Activities of children's playing occur in the daytime, and adults' exercise occur in the night time, that is, space has different kinds of activities and user groups during day and night |
| Happy | 78 | Children can enjoy playing a variety of facilities in this park<br>Children can make new friends in this park |
| Comfort | 72 | Enough sitting space |
| Open/Good views | 53 | Open space, which has been suitable for human-scale activities<br>Adults can easily keep their eyes on their children playing nearby due to the reasonable viewing distance |

# Appendix F: Questionnaire Topic Guide

This topic guide is used for the street-intercept questionnaire with residents living in the SSTEC. The purpose of these questions aims to obtain people's more detailed perspectives and objective descriptions of those 'invisible' needs towards public places and social facilities within the eco-city.

Dear Interviewee,

The main purpose of this questionnaire is to understand people's attitudes towards the current situation of public space and people's needs when staying in public space. The anonymous method of random sampling is adopted, and all results of the questionnaire are only used for the academic research. There are no right or wrong answers and options, please read carefully and answer according to the actual situation. This questionnaire will take about 10–15 minutes of your time.

The public space involved in this questionnaire refers to those outdoor spaces that people use in their daily lives, including streets, squares, outdoor spaces in residential areas, parks and playgrounds.

**Part 1: Basic Information**

1. Your gender: () Male; () Female
2. Your age: () 16-24 years old; () 25-40 years old
() 41-60 years old; () 60 years old and above



Y. Wu, *Possibilities and Limitations of Chinese Eco-City Development*, New Perspectives on Chinese Politics and Society,
https://doi.org/10.1007/978-981-95-1958-3

3. Your occupation:
() students; () office workers;() retirees; () freelancers

## Part 2: People's Outdoor Activities in the Public Space

5. What is the main purpose of coming here for you? What outdoor activities do you usually do? (such as fitness, play, chat with neighbours)

____________________________________________

6. How can you come here? (multiple choices are allowed) And do you think is it convenient for you to come here?
() walking; () bicycle; () subway; () bus; () taxi; () private car
() It is convenient. () It is not convenient.
7. What is the usually time and frequency of your activities here? (multiple choices are allowed)
Time: () morning on weekdays; () afternoon on weekdays;
() weekend morning; () weekend afternoon
Frequency: () daily; () weekly; () several times a month
() rarely (about several times a year)
8. Would you like to spend more time here enjoying the outdoor space?
() Yes, I would like; () No, I would not like.

## Part 3: Spatial Quality and Facilities

9. What aspects of the environment here do you think need to be improved? (multiple choices are allowed)
() can easily get access (e.g. convenient parking spaces, more accessible walking space); () safe and enjoyable outdoor space
10. What aspects of facilities do you think need to be improved? (multiple choices are allowed)
() Quantity and quality of fitness equipment;
() Facility corridors for shelter from wind and rain;
() Barrier-free facilities (referring to the service facilities provided to ensure the safety and convenience of the disabled, the elderly, pregnant women, children, also referring to toilets for the disabled).

## Part 4. Walking Trails (The Space for Pedestrians and Cyclists)

11. What walking trails within the SSTEC are you often use? (multiple choices are allowed)
() Pedestrian or bicycle paths on main and secondary roads (such as Zhongxin Avenue and Central Avenue);
() Eco Valley; () Huifengxi Greenway; () Ganluxi Greenway; () Nandi or Dongdi Walking Trails
12. What is your main purpose of using the Eco-City walking trail? (multiple choices are allowed)
() Commuting to and from get off work or taking children to and from school;
() Running, walking, cycling, fitness;
() Passing by when going to parks, supermarkets, restaurants, community centres;
() Others: ______________________________
13. What do you think the main problems of the current walking trails within the SSTEC?
() Unsafe: lack of protective measures from the motorway (such as green belt isolation) Road name: ____________________
() Unobstructed: walking trails are often occupied by motor vehicle parking. Road name: ____________________
() Inconvenient: it is difficult to find shared bicycles and suitable parking spaces. Road name: ____________________
() Uncomfortable: the walking trails is too narrow or lack of shade Road name: ______________________________
() Unattractive: lack of recreational features (such as lack of running tracks or fitness facilities)
Road name: ______________________________
() No problem, basically satisfied.

## Part 5. Overall Attitudes Towards the Spatial Quality (Only Referring to This Research Location)

14. Besides good and bad, can you please write three adjectives to describe the overall spatial environment?

# Appendix G: Interview Topic Guides

- This topic guide is to be used in interviews with planning practitioners who are related to the SSTEC project.
- The purpose of these interview questions has been seen as gaining the insightful understanding on the formation and forms of *humanistic planning under the context of ecology*, as well as the factors that promote and limit people-oriented concept implemented in the Chinese eco-city planning project.
- Interviewees would be classified into three groups, all related to the SSTEC project—*official of Tianjin Planning Bureau*, *official of House and Urban-Rural Development Bureau*, *member of China-Singapore Tianjin Eco-City Administrative Committee.*
- Interviewers are all explained the following precautions:

   (1) Notes about key information will be taken by the interviewer, and audio recording will be used if permitted by the interviewee.
   (2) Interviewees' responses will be confidential. The researcher will not name anyone in this research. Also, all personal information will not be passed onto any third parties.
   (3) All interviewees will be provided with the opportunity to remain completely anonymous when mentioned in the book. In addition, all data will be stored securely, and all personal information will not be passed onto any third parties (detailed



Y. Wu, *Possibilities and Limitations of Chinese Eco-City Development*, New Perspectives on Chinese Politics and Society,
https://doi.org/10.1007/978-981-95-1958-3

measures are shown in Appendix X. Participant Information Sheet).

**Part 1. Interviewees' Personal Information**

What department are you in? What is the main duty and routine of your work related to the SSTEC project?

**Part 2. People-Oriented Planning Concept and Its Future Trends in Eco-City Development**

2.1 How do you understand the concept of '*humanistic planning under the context of ecology*' and the concept of '*harmonious people-place relationship*'?
2.2 Does the SSTEC produce any humanistic planning strategy or plan in the policy-making or index-making processes?

**Part 3. Documentary Analysis and Questionnaire Findings**

3.1 What is the most important driver of people-oriented planning to be implemented in the eco-city development do you think and why?

- Planning documents show that the main driver of humanistic planning in the context of the ecology is 'the intention of national government for ecological civilization construction'. Do you agree? And are there other drivers for the SSTEC to promote the people-oriented planning concept?
- In your practical experience, in what ways does the SSTEC become the pilot project for other Chinese cities to promote the humanistic planning concept?
- In your practical experience, how does the SSTEC respond to the transformation of its development path from the previous economy-oriented (KPI-driven) to the current people-oriented pattern?
- Based on your answer, to what extent do you think the people-oriented planning principle will replace economy-oriented concept to become the main driver in the future for all Chinese eco-city projects?

3.2 How planners/designers consider people's needs during policy-making or place-making stages? For example, in the questionnaire for eco-city's residents and my personal field observation, I found that most people talked about the safety issues as the primary consideration of their sense of activity experience. Can I ask whether designers consider safety issues in the first place during the place-making process?

**Part 4. Challenges of Humanistic Planning in the Eco-City Development and Approaches to Overcome**

4.1 In your practical experiences, what are biggest differences between traditional Chinese cities' development pattern and the people-oriented newly built eco-city? (esp. in different stages: planning, construction and management)
4.2 To what extent do you want to change your current traditional planning strategies and design skills to build the built environment characterized as *harmonious people-place relationship*?
4.3 In practice, what is the most serious difficulty you meet when developing people-oriented eco-city project? How do you handle them? What changes you made as a result?

**Part 5. Research Questions for Supplementary Fieldwork**

5.1 How do you think SSTEC has changed in its aims and strategic focus since its inception, and why?
5.2 Is SSTEC still an eco-city, or is it now more of a smart and/or digital city in focus?
5.3 Were there any other features including policy or planning strategic that were taken from Singapore? How Singaporean Government or planning authorities were involved or were they involved in any way?

# Appendix H: Participant Information Sheet

**Part 1. Participant Information Sheet (For Questionnaires)**

1. **Research Project Title**
Possibilities and Limitations of Chinese Eco-City Development: Case Study of Sino-Singapore Tianjin Eco-City.
2. **Invitation**
You are being invited to take part in this research project. Before you decide to do so, it is important for you to understand why this research is being done and what it will involve. Please take time to read the following information carefully. Ask me if there is anything that is not clear or if you would like more information. Take time to decide whether or not you wish to take part. Thanks for reading this.
3. **Why Me? Why Should I Take Part?**
It would not be practical for me to gather information from everyone, so to ensure fairness, I randomly select a number of people living/working/visiting Tianjin eco-city to help me complete this questionnaire. I involve you as I value your contribution and listen to your voices.



Y. Wu, *Possibilities and Limitations of Chinese Eco-City Development*, New Perspectives on Chinese Politics and Society,
https://doi.org/10.1007/978-981-95-1958-3

This study aims to provide the up-to-date information about your needs of social facilities and your outdoor activities in public spaces. The information collected will help to provide both planning practitioners and management committee with a clear picture of the communities' views, and in turn, help to improve your living environment and living quality.

4. **What Is the Project's Purpose?**

The main objective of this research is to argue Chinese eco-city planning should be deemed as the humanistic planning, and its goal is to build sustainable and liveable habitat for people by taking their daily behaviours and needs into account. To achieve this goal, considering the interactive relationship between places (public spaces) and people (people's activities) during planning process is an innovative way to improve Chinese eco-city's spatial quality. As this research project is based upon the case study of Tianjin eco-city, the main objective of field trip in China is to explore people's public lives in public spaces within Tianjin eco-city.

5. **Do I Have to Take Part?**

It is up to you to decide whether or not to take part. If you do decide to take part you will be able to keep a copy of this information sheet and you need to indicate your agreement to the consent form. You can still withdraw at any time. You do not have to give a reason.

6. **What Will Happen to Me if I Take Part?**

You will be asked to complete a paper-based questionnaire which takes you about 15 minutes to answer or will be observed about what activities you do in public spaces. You may also wish to agree to a follow-up interview to find out more about your approach.

7. **What Are the Possible Disadvantages and Risks of Taking Part?**

Participating in the research is not anticipated to cause you any disadvantages or discomfort. The potential physical and/or psychological harm or distress will be the same as any experienced in everyday life.

8. **What Are the Possible Benefits of Taking Part?**

While there are no immediate benefits for those people participating in the project, it is hoped that this work will have a beneficial impact on improving the living environment of eco-city based on your voices.

**9. Will I Be Recorded and How Will the Recorded Media Be Used?**
You will not be recorded in any way other than your input to the questionnaire without separate permission being gained from you.

**10. Will My Information Be Kept Safe?**
Your information will be treated in the strictest confidence. Any published statistics will not identify any individual or business. I recognize that you are sharing confidential/personal information with us and treat it as confidential. Keeping information secure is a fundamental part of School of Geography and Planning, Cardiff University.

This research will receive full ethical approval from School of Geography and Planning, Cardiff University. And it will strictly uphold the data protection protocols in accordance with the Data Protection Act. All Participants will be provided with the opportunity to remain completely anonymous, and personal information will be not be passed onto any third parties.

**11. How About the Data Management?**
All data collected by the researcher will be stored in a locked file saved in the researcher's computer, only the researcher knows the locked password and this file will not be shown to anyone else. As data analysis will last about two years, and book writing will last within one year, thus, all data will be kept within three years, that is, they will be destroyed after the book writing completed, normally by August 2022. In addition, the researcher promise that all data collected in China will not be shown to anyone and any third parties.

**12. How Can I Get in Contact with You?**
If you have any questions or concerns regarding the research request, please feel free to contact me at any time.

## Informed Consent (For Questionnaire)

This is a consent sheet that includes all of the required information that research participants are required to know before giving consent. Please read this consent document carefully before you decide to participate in this study.

**Research Title:** Possibilities and Limitations of Chinese Eco-City Planning: Case Study of Sino-Singapore Tianjin Eco-City.
**Research Purpose:** To explore whether the interactive relationship between people and place is an innovative way for Chinese eco-city development, by exploring people's public lives in public spaces within Tianjin eco-city.
**Who is conducting and funding this research:** Yani Wu, PhD student of Cardiff University, is conducting this research and is funding this project herself.
**What you will be asked to do in this study:** To complete the paper-based questionnaire which takes you 15 minutes to answer or will be observed about what activities you do in public spaces.
**Confidentiality:** Your identity will be kept confidential to the extent provided by law. Your information will be assigned a code number that is unique to this study. The list connecting your name to this number will be kept in a locked file and only the researcher Yani Wu and her supervisor Dr. Patricia Aelbrecht can see the questionnaire you participated in. No one in China will be able to see your questionnaire or even know whether you participated in this study. Study findings will be presented only in summary form and your name will not be used in any report.
**Anonymity:** In order to safeguard the participants, every person in this study, and most of the scenes including the name of living places, are carefully protected by anonymity. Your identity in this study would also be anonymous. It will not be possible to know who chooses to participate in this study and who did not. Material will not be published in China.
**Right to withdraw from the study:** You can choose to stop participating at any time.

**You will be given a copy of this form whether or not you agree to participate.**

**If you agree to participate in this study, please sign below. Thank you.**

**Agreement:**

I have read the procedure described above. I voluntarily agree to participate in the procedure and I have received a copy of this description.

**Name (Printed):** ________________________________
**Signature:** ________________________________
**Date:** ________________________________
**Principal Investigator:** ______________________
**Date:** ____________________________

**Part 2. Participant Information Sheet (For Semi-Structured Interviews)**

1. **Research Project Title**
Possibilities and Limitations of Chinese Eco-City Development: Case Study of Sino-Singapore Tianjin Eco-City.
2. **Invitation**
You are being invited to take part in this research project. Before you decide to do so, it is important for you to understand why this research is being done and what it will involve. Please take time to read the following information carefully. Ask me if there is anything that is not clear or if you would like more information. Take time to decide whether or not you wish to take part. Thanks for reading this.
3. **Why Me? Why Should I Take Part?**
It would not be practical for me to gather information from everyone, so to ensure fairness, I randomly select a number of people involved in planning and design process of Tianjin eco-city to help me complete this interview. I involve you as I value your contribution and listen to your voices.
This study aims to provide the up-to-date information about design principles of spatial environment based on people-centred planning concept in your daily work. The information collected will help this research to identify whether considering people's needs and behaviours is an innovative way for Chinese eco-city development.
4. **What Is the Project's Purpose?**
The main objective of this research is to argue Chinese eco-city planning should be deemed as the humanistic planning, and its goal is to build sustainable and liveable habitat for people by taking their daily behaviours and needs into account. To achieve this goal, considering

the interactive relationship between places (public spaces) and people (people's activities) during planning process is an innovative way to improve Chinese eco-city's spatial quality. As this research project is based upon the case study of Tianjin eco-city, the main objective of field trip in China is to explore planners' insightful understanding about the humanistic planning concept under the context of ecology.

5. **Do I Have to Take Part?**

It is up to you to decide whether or not to take part. If you do decide to take part you will be able to keep a copy of this information sheet and you need to indicate your agreement to the consent form. You can still withdraw at any time. You do not have to give a reason.

6. **What Will Happen to Me if I Take Part?**

You will be asked to conduct the semi-structured interview which takes you about 40 minutes, discussing about your work and view-points about people-centred planning concept in Tianjin eco-city's planning process.

7. **What Are the Possible Disadvantages and Risks of Taking Part?**

Participating in the research is not anticipated to cause you any disadvantages or discomfort. Potential physical and/or psychological harm or distress will be the same as any experienced in everyday life.

8. **What Are the Possible Benefits of Taking Part?**

While there are no immediate benefits for those people participating in the project, it is hoped that this work will have a beneficial impact on improving the living environment of eco-city based on your voices.

9. **Will I Be Recorded and How Will the Recorded Media Be Used?**

You will not be recorded in any way other than your input to the interview without separate permission being gained from you.

10. **Will My Information Be Kept Safe?**

Your (institution) information will be treated in the strictest confidence. Any published statistics will not identify any individual or business. I recognize that you are sharing confidential/personal information with us and treat it as confidential. Keeping information secure is a fundamental part of School of Geography and Planning, Cardiff University.

This research will receive full ethical approval from School of Geography and Planning, Cardiff University. And it will strictly uphold the data protection protocols in accordance with the Data Protection Act. All Participants will be provided with the opportunity to remain completely anonymous, and personal information will be not be passed onto any third parties.

11. **How About the Data Management?**

All data collected by the researcher will be stored in a locked file saved in the researcher's computer, only the researcher knows the locked password and this file will not be shown to anyone else. As data analysis will last about two years, and book writing will last within one year, thus, all data will be kept within three years, that is, they will be destroyed after the book writing completed, normally by August 2022. In addition, the researcher promise that all data collected in China will not be shown to anyone and any third parties.

## Informed Consent (For Semi-Structured Interviews)

This is a consent sheet that includes all of the required information that research participants are required to know before giving consent. Please read this consent document carefully before you decide to participate in this study.

**Research Title:** Possibilities and Limitations of Chinese Eco-City Development: Case Study of Sino-Singapore Tianjin Eco-City.

**Research Purpose:** To explore whether the interactive relationship between people and place is an innovative way for Chinese eco-city development, by exploring planners' insightful understanding about the 'humanistic planning under the context of ecology' concept.

**Who is conducting and funding this research:** Yani Wu, PhD student of Cardiff University, is conducting this research and is funding this project herself.

**What you will be asked to do in this study:** To conduct a semi-structured interview which takes you about 40 minutes.

**Confidentiality:** Your identity will be kept confidential to the extent provided by law. Your information will be assigned a code number that is unique to this study. The list connecting your name to this number will be kept in a locked file and only the researcher Yani Wu and her supervisor Dr. Patricia Lopes Simoes Aelbrecht can see the interview transcript you participated in. No one in China will be able to see your interview transcript or even know whether you participated in this study. Study findings will be presented only in summary form and your name will not be used in any report.

**Anonymity:** In order to safeguard the participants, every person in this study, and most of the scenes including your work institution and job title, are carefully protected by anonymity. Your identity in this study would also be anonymous. It will not be possible to know who chooses to participate in this study and who did not. Material will not be published in China.

**Who to contact if you have questions about this study:**

Yani Wu, Ph.D. student, School of Geography and Planning, Cardiff University.

**You will be given a copy of this form whether or not you agree to participate.**

**If you agree to participate in this study, please sign below. Thank you.**

**Agreement:**

I have read the procedure described above. I voluntarily agree to participate in the procedure and I have received a copy of this description.

**Name (Printed):** ______________________________

**Signature:** ______________________________

**Date:** ______________________________

**Principal Investigator:** ______________________

**Date:** ______________________________

# Appendix I: Location in Relation to Notable Destinations




Y. Wu, *Possibilities and Limitations of Chinese Eco-City Development*, New Perspectives on Chinese Politics and Society,
https://doi.org/10.1007/978-981-95-1958-3

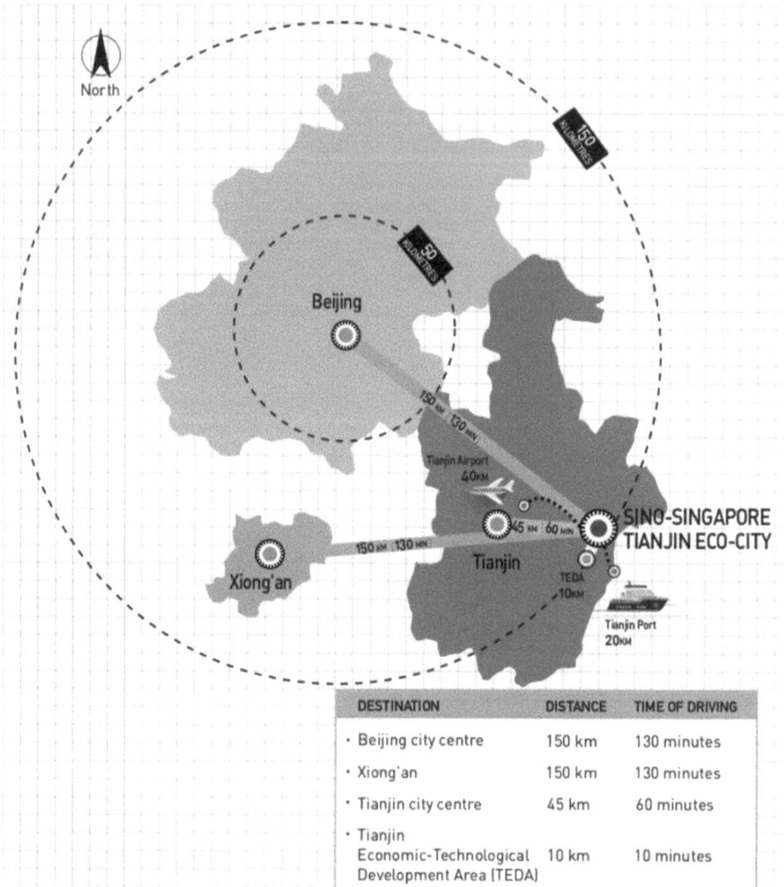


| DESTINATION | DISTANCE | TIME OF DRIVING |
|---|---|---|
| • Beijing city centre | 150 km | 130 minutes |
| • Xiong'an | 150 km | 130 minutes |
| • Tianjin city centre | 45 km | 60 minutes |
| • Tianjin Economic-Technological Development Area (TEDA) | 10 km | 10 minutes |
| • Tianjin Airport | 40 km | 45 minutes |
| • Tianjin Port | 20 km | 30 minutes |

(Source: Map is from Sino-Singapore Tianjin Eco-City Administrative Committee et al., 2021. p. 25.)

# Appendix J: Summary of Research Design

This research starts with the rationale for choosing the case study strategy, then it demonstrates specific research methods relating to the selected case study. Following the research strategy characterized as three-layer research design with a single case study, research methods include document analysis, *Public Space and Public Life (PSPL) Survey* and in-depth qualitative interviews. In addition, this chapter also discusses some limitations of chosen research methods and several potential ethical issues.

## Overall Information About Research Design and Data Description

### *Methodological Choice*

This research situates in *the humanistic planning under the context of ecology*, an applied field that requires insights from multiple disciplines—urban planning, urban design and psychology respectively. In a narrow sense, it intends to explore people's behaviours in public spaces within the eco-city, in order to search for the interaction between people and place. From the perspective of research aim listed in Chapter 1, it requires qualitative data and suggests the necessity of researcher's involvement as well as the (participatory) observation. In a broad sense, this research is expected



Y. Wu, *Possibilities and Limitations of Chinese Eco-City Development*, New Perspectives on Chinese Politics and Society, https://doi.org/10.1007/978-981-95-1958-3

to discuss some specific challenges, issues or problems of humanistic planning and to deepen (and also broaden to a small extent) the implication of people-place interactive relationship in Chinese eco-city development. As qualitative method focuses on the in-depth validity over generalizability, data collection and analysis used by qualitative techniques are drawn from particular context in this research.

This research has adopted a holistic-view research approach, that is, both inductive (data-driven) and deductive (theory-driven) analysis (Braun et al., 2019, p. 297) are combined to interpret qualitative data collected from the case study. The reason for this choice is to maximally find out the match (applicability) between global universal theories and eastern practices for ensuring the result of this study is trustworthy and meaningful, as there is a certain degree of difference between global universal theories in guiding eastern practices. Specifically, *the humanistic planning concept under the ecology context* has been seen as the focal point of this research, in Chinese context, the appropriate theory might be the traditional concept of *the unity of man and nature* (天人合一), this research aims to learn from this interactive relationship, to see whether there are any evidences to show global universal humanistic planning and design concept has happened in Chinese eco-city development, if yes, in what ways and to what extent. One the one hand, from the perspective of inductive analysis, such analytical method can capture both manifest and latent meanings (Braun et al., 2019, p. 851), as it not only summarizes and reorganizes observational diaries, questionnaires and interview transcripts, more importantly, it indeed contributes to connecting research objectives with research questions to identify and to interpret key features of research data. On the other hand, from the perspective of deductive analysis, this research innovatively combines both humanistic and ecological elements from different disciplines, in order to build the fundamental conceptual and analytical framework for guiding the whole research process. Focusing on *the humanistic planning under the context of ecology*, *the people-place interactive relationship*, derived from three theories in different disciplines has served as the core connotation of people-oriented concept, which is used to guide data collection during the fieldwork and data analysis after the fieldwork. Specifically, the *Public Space and Public Life (PSPL) Survey* is used to collect data about residents' outdoor activities (people's behaviours) in public spaces within the eco-city (place) and their attitudes or opinions (people's cognition) on the spatial quality of public spaces within the eco-city (place). Moreover, data discussion has always revolved around the people-place relationship derived from the combined theoretical framework, to explore both forms and implications of behavioural connection between people and place in the eco-city development, and also to response to *Objective Three*.

### *Details of the Type and Amount of Data*

Firstly, the questionnaire survey was carried out through printed documents, being distributed at four research locations. To achieve an effective response rate and to get more residents' engagement, approximately 100 questionnaires are planned to be distributed at each research site. 400 questionnaires are returned with 14 incomplete questionnaires, 386 are valid. The final response rate is 96.5%, which is acceptable as it is representative by living residents of different ages. Also, high frequency word analysis is used to identify the most frequently occurring adjective words, receiving respondents' insights into prevalent themes about their overall attitudes towards the spatial quality.

**Summary of Questionnaire Survey Information**

| *Research location* | *Number of questionnaire* | *Valid questionnaire* | *Response rate (%)* |
|---|---|---|---|
| Zhongxin Avenue | 80 | 75 | 93.75 |
| 3rd community centre | 115 | 112 | 97.4 |
| Eco-valley | 120 | 117 | 97.5 |
| Children's park | 85 | 82 | 96.5 |
| Total Number of Questionnaire: 80 + 115 + 120 + 85 = 400 | | | |
| Total Number of Valid Questionnaire: 75 + 112 + 117 + 82 = 386 | | | |
| Total Response Rate: 386 / 400 *100% = 96.5% | | | |

Secondly, each street-intercept interview with residents lasts appropriately half an hour, and section criteria of representative interviews is seen as exploring residents' positive and negative attitudes towards spatial elements and factors that could affect the spatial quality of public spaces within the SSTEC. Based on such criteria, two representative street-intercept interviews conducted in Zhongxin Avenue have been selected among five ones to be further analysed, six representative street-intercept interviews conducted in the 3rd community centre have been selected among ten ones, four representative street-intercept interviews conducted in Children's Park have been selected among five ones, and two representative street-intercept interviews conducted in the 4th community centre have been selected among four ones.

Thirdly, approaches of participation observation include mapping, on-site counting, photography and diary keeping. Mapping and on-site counting are conducted simultaneously, aiming to record the pedestrian flow of four research locations with two-hours recording time as a slot. Detailed description is shown in *Appendix E*, and data analysis used by the fixity-flow attachment framework will be further analysed in Chapter 6. And final results about traffic flows are not only used to describe place attachment but also to show the spatial quality of public spaces.

Lastly, in-depth interviews with related planning practitioners of the SSTEC concern the embodiment of people-oriented concept in the SSTEC project, using this case study to respond to research question two—how Chinese eco-city planning regulations and proposed projects consider people-oriented concept current and future eco-city development.

## Summary of In-Depth Interviews

| *Interviewee and company* | *Interview time* | *Format* |
|---|---|---|
| Member of Tianjin Eco-city administrative committee | 20 July 2019 | Face-to-face |
| Official of Tianjin Planning Bureau<br>Member of SSTEC group | 10 March 2021 | Online |
| Deputy director of Sino-Singapore eco-city construction bureau | 16 May 2023 | Face-to-face |
| General leader of L&A design | 16 April 2023 | Face-to-face |

# The Three-Layer Research Design with a Single Case Study

For a more logical and systematic data collection and analysis, *the three-layer methodological framework with a single case study* has been designed as the research strategy for this research, for exploring research data about the people-place interactive relationship in Chinese eco-city project, and for examining people's needs and behaviours when they are staying in public spaces. A set of qualitative research methods are properly chosen and are used in each data collection procedure. Three influencing factors have been considered for such selection of research strategy—the type of research questions, the degree of control over actual behavioural events faced by the researcher and the focus of attention about contemporary phenomena (Yin, 2014).

In this research, the typical qualitative methods, such as participant observation, semi-structured interviews and qualitative self-completion questionnaire guarantee 'complete answers to research questions and balanced results' (Johnson et al., 2007, p. 125). It is worth mentioning that data collection under such strategy has been designed as the so-called *observation sandwich* to ensure the reliability of data, and to increase credits of the internal validity. That is, verification becomes the theme of this research strategy—participant observation used during the pilot study is followed by the street-intercept survey, and then they are followed by in-depth interviews, with the single case study design acting as the toothpick of *observation sandwich.*

## *Single Case Study Strategy*

Case study refers to a programme or an activity framed in the specific time and place. According to Bhattacherjee (2012), a case study is a suitable choice for research, where the phenomenon during a period of time in one or several sites can be intensively researched. It is 'defined by interest in an individual case, not by methods of inquiry used, and the object of study is a specific, unique and bounded system' (Stake, 2008, p. 123). Case study helps the researcher to understand 'complex, temporal processes rather than factors or causes' (Bhattacherjee, 2012, p. 94), and contributes to providing more contextualized and more authentic interpretation of the phenomenon.

Yin (2014) demonstrated both importance and applicability of case study design in the explanatory research of social science. It is functioned as an empirical inquiry that can research contemporary social phenomenon in depth within the real-world context, especially under the situation of unclear boundaries between social phenomenon and context. The suitability of case study is seen as 'when how and why questions being asked, the subject of the research is a contemporary event, and the researcher has little control over relevant behaviours involved in the event' (pp. 15–16).

The overall aim of this research focuses on the understanding about humanistic issues that happened in Chinese eco-city development, instead of the comparative or general perception about eco-city planning. Specially, this research strategy is designed based upon research questions to show how research data has been collected and analysed from two aspects—on the one hand, how to define the people-place interactive relationship (how do people measure the spatial environment by their behaviours and how does the spatial environment impact people's behaviours), and how this interactive relationship influences eco-city development (whether the design pattern considering people-place relationship can be regarded as the possible improved approach to Chinese eco-city development). On the other hand, what are constraints and initiatives faced by Chinese eco-city development when implementing the humanistic planning concept. Based on the above two points, there has been more than one variable of interest (e.g. people, place and constraints) and multiple sources of needed evidence in this research, the researcher has less control.

Such designed case study strategy makes a contribution to this research, and also to existing literatures, especially due to the lack of discussion about people-oriented planning concept under the context of ecological development in academia. Thus, more exploration and expanding on recent eco-city's academic research need to be explicitly made by a broader humanistic perspective and by the interdisciplinary way of thinking. That is, this research design aims to understand topics and issues of humanistic planning concept from multiple theories in literatures and to refine the people-place relationship from the chosen case study. In addition, referring to the category of case study, Yin (2014) classified three types—exploratory, explanatory and descriptive ones, which are not seen as being separate or being hierarchy. This research is designed as the *exploratory-explanatory* case study to make research findings be more convinced.

As Flyvbjerg (2006) illustrates, the successful choice of paradigmatic case cannot be guaranteed in advance of the study itself. 'There was no assumption, therefore, that the case would, on closer examination, reveal itself to be truly paradigmatic, rather than force a categorization on the case through which to interpret the findings, a more grounded approach was adopted, whereby the case was expected to be uniquely shaped by its particular context' (Robert, 2016, p. 131). *The Sino-Singapore Tianjin Eco-City (SSTEC)* has been chosen as a case to examine and to refine the analytical framework by exploring some typical issues and concerns about the humanistic planning and design concept (people-place relationship), which does not exist in literatures. Further crucial reason for selecting the SSTEC as the single case study to research is because it appears to represent what Flyvbjerg (2006) believes a 'critical case'. As mentioned above, a case study design has been adopted to answer research questions and to collect data for exploring real-world situations about the interactive relationship pattern between people and place within Chinese eco-city. Process of choosing a case study sample is selective with a purpose. Accordingly, a research sample should be selected to investigate the people-oriented planning concept (public life in public spaces) in full details. Thus, choice of a case study sample tends to be the most influential and representative Chinese eco-city project to the application of humanistic planning and design concept. That is, the chosen sample ought to obviously represent the prevailing planning practices and also it may potentially influence the future development of Chinese eco-cities.

Caprotti and Gong (2017) made the case for 'humanizing' newly built eco-cities by emphasizing on urban social sustainability through the lens of experiences of new residents. They argued 'there is a need to focus on the ways in which socially sustainable urban environments are constructed, in new urban spaces, through relational networks comprised by interactions between residents, buildings, facilities and specific (e.g. domestic) spaces' (p. 45). However, as Joss and Molella (2013) revealed in their research, 'Chinese urban experiment as a *techno-city*, putting much emphasis on how the eco-city development is carried out in conjunction with the development of a broader technological project' (p. 121). Moreover, current academic studies on the SSTEC, the first eco-city that developed collaboratively between countries and the experimental and innovative pilot project in China, mainly focus on the descriptive or technological analysis about planning documents and second-hand planning policies. They examine 'whether eco-cities exemplify urban sustainability ideals, discuss challenges of implementing eco-cities, or question how eco-city's experience can be used to develop indices and assessment measure for future projects' (Chang, 2018, p. 62); lack of research focuses on the humanistic aspect and social sustainability. Such research gap in academic literatures has determined the reason for case study selection—combining the humanistic planning and design concept with the SSTEC practice.

### *The First Layer: An Exploratory Pilot Study*

In order to investigate whether crucial components of this research are feasible and can predict an appropriate sample size for the full-scale research, a small-scale pilot study with an exploratory case study has been adopted as the first layer of the research design. The most important reason for conducting the pilot study is because it is crucial for the researcher to avoid wasting time and resources that can collect enough data that responds to answering research questions.

The first layer of 'sandwich' is made of desk research and preliminary participant observation by the researcher known as the pilot study. The aim of this stage is to broaden more knowledge on research topics and to explore the initial information about the SSTEC project, checking the sampling and evaluating the research design. During this stage, it aims to narrow down research questions from the analytical framework and to identify how research questions are properly formulated for the application to research methods. It is worth mentioning that document analysis and preliminary participant observation at this stage respond to *Objective Two* and *Research Question One*, by reviewing core influencing factors for supporting Chinese eco-city development as well as how Chinese urban development move to the new development stage of people-oriented eco-city.

Key research questions guiding this initial phase of research are stated below. Two main questions, shown as follows, are discussed in more detail in the conclusion section of Chapter 7, in the light of preliminary document analysis and interviews.

> *What are limitations and challenges of Chinese eco-city development? (KPI/Technology-Driven)*

*What are innovative initiatives to make people-oriented eco-city during the current development process of smart/digital city?*
*How does the 'planning for people' concept in Chinese eco-city project (SSTEC) differ from the conceptualization in official documentation?*
*Are there any evidence to show such planning and design concept already accepted by the means of planning requirements?*

The approach in this stage includes three elements:

① Desk research into the broader social context of the SSTEC case
② Reviews of official documents
③ Observation research in public spaces within the SSTEC (1st fieldwork)

(1) Purposive Sampling

The logic of purposeful sampling depends on selecting information-rich case for the in-depth study, which can learn much about 'issues of central importance to the purpose of research' (Patton, 2015, p. 170). Sampling within a case plays a vital role in the research design, where 'the case is not so small that it can be subjected to exhaustive investigation' (Atkinson & Hammersley, 2007, p. 35). Full considerations need to be taken into account about when and where to observe, who to interview and what to ask, as well as how to record research data for further analysis. Atkinson and Hammersley (2007) proposed three major dimensions of sampling within the case study, namely time, people and context, which will be further discussed in the following section. It is important to attempt to ensure research data generated from the case be adequately sampled, and to reflect on any inadvertent sampling that has taken place, by 'making any criteria employed as explicit and as systematic as possible' (Atkinson & Hammersley., 2007, p. 35).

Purposive sampling, rather than random sampling, has been seen as the appropriate method adopting in the explanatory case study. It is based upon the guideline of theoretical sampling—the selected subjects have shared key characteristics that the researcher emphasizes in his/her research (Yin, 2014).

### a. (Observing) Time

People's activities frequently change over time in different ways, even within the same locality. For example, a community park in the day time could be seen as the place for the elderly to rest and to chat, but in the night time could be used as the place for the young to exercise or to gather together. From this perspective, time has been seen as one of significant dimensions, which cannot be neglected in the observational research. 'Any attempt to represent the entire range of persons and events in the case under study will have to be based on adequate coverage of temporal variation' (Atkinson & Hammersley, 2007, p. 36.). As is known, it is hard for the researcher himself/herself to conduct fieldwork around the clock, 'long bursts of observation, uninterrupted by periods of reflexive recording, will thus tend to result in data of poor quality' (*ibid*). In other words, the researcher needs to resist temptation about observing or recording everything that goes on during the fieldwork. A more selective method contributes to collecting data of better quality, by providing several time periods of observation. As Schatzman and Strauss demonstrated (1973),

> If the researcher elects to observe work around the clock, he can first observe a day shift for several days, then evenings and then nights, for a period of consecutive days until he is reasonably familiar with all three shifts. Or he may cover events at any given sub-site by overlapping time on consecutive dates—for example, 7 am–9 am, 8 am–10 am, 9 am–11 am—and over a period of days cover the organization around the clock. And similar considerations to those outlined above will also apply to larger-scale temporal dimension, such as seasonable or annual cycles (p. 39).

It is worth mentioning that even though such selective method sounds more related to physical sciences-style or biosciences-style research, it is relevant to urban studies as it provides a comprehensive view of the intricate dynamics and changes within urban environments. By capturing short-term fluctuations and long-term trends, researchers could learn more about how people behave for addressing multifaceted challenges of urban living.

Considering time and physical energy of the researcher, data collection plans to conduct in two periods of time—winter and summer, for checking observation results and further ensuring the reliability of research data. Reason for this arrangement is mainly because in the pilot study, especially for the first several days (18 December 2018 and 19 December 2018) in the SSTEC, few people did outdoor activities in the winter due to the cold temperature. Another reason is because, after each observation for the specific research locations within the SSTEC, the researcher kept the following questions in mind—'what is the time frame surrounding the behaviour as it naturally occurs in the environment? Does the behaviour begin repeating itself every twenty-four-hour period or other fixed time?' (Bechtel et al., 1987, p. 16).

Being the observer, the researcher took notes and mapped people's activity during the participant observation period, by using the pre-prepared site map and template to record data. Moreover, the fixed time slot has been adopted in order not to presuppose people's activities and patterns for using social infrastructures. Time-frame defined as certain hours of the day, consisting of observations in the morning (8am–10am) when people go to work at the peak hour, at noon (1 pm–3 pm) when people have lunch outside and in the evening (6 pm–8 pm) when people go home after work and take outdoor exercise. Two days were spent per research site, one working day and one weekend respectively, during a total six-week period (three weeks in winter and three weeks in summer) with the good 'time-out' conditions for participant observation. In short, this time strategy allows a good cross comparison about people's outdoor activities in different seasons, which can help to give the researcher a real picture of the public lives within the eco-city. The following two charts present timeframe and specific dates of *Public Space and Public Life Survey (PSPL)* data collection.

## Time Horizon and Condition

| *Time condition* | *Time horizon* |
|---|---|
| Seasons when the weather is suitable for travel (sunny and cloudy days) | Winter (December2018)<br>Summer (July 2019 and June 2023) |
| Time covers working days and weekends | At least two recording times for each working day and weekend in each research site |
| 'Time-out' hours | Weekdays: lunchtime activity peak (1 pm–3 pm) and after working hours (6 pm–8 pm);<br>Weekends: (10 am–8 pm) when retail activities are open |

(continued)

(continued)

| *Time condition* | *Time horizon* |
|---|---|
| Peak-hours | Mornings (8 am–10 am) and Sunset time (6 pm–8 pm): more people taking buses going to or coming from work, and doing exercise after all-day work |

## Specific Dates of Data Collection

| *Research site* | *Time of field work* |
|---|---|
| Zhongxin Avenue | 20 December 2018 (Thursday)<br>22 December 2018 (Saturday)<br>25 June (Sunday) and 26 June 2023 (Monday) |
| The 3rd community centre | 17 July 2019 (Wednesday)<br>20 July 2019 (Saturday)<br>25 June (Sunday) and 26 June 2023 (Monday) |
| The 4 community centre (additional research site) | 25 June (Sunday) and 26 June 2023 (Monday) |
| Eco-valley<br>(selected community section) | 13 July 2019 (Saturday)<br>15 July 2019 (Monday)<br>25 June (Sunday) and 26 June 2023 (Monday) |
| Children's park | 14 July 2019 (Sunday)<br>16 July 2019 (Tuesday)<br>25 June (Sunday) and 26 June 2023 (Monday) |
| Each research site must ensure to coverage both weekday and weekend | |

(Source: Information is summarized by the researcher.)

b. **(Observation) People**

The sample of people in this research has been divided based on the particular context of each research question. The criteria of participants selection falls in two categories—residents of eco-city based on research question two presented in Chapter 6, and planning practitioners related to the SSTE project based on research question three presented in Chapter 7.

### c. **Context (Observing Place)**

Taking account of variations in context is as important as sampling across time and people. 'Within any setting people may distinguish between a number of quite different contexts that require different kinds of behaviours' (Atkinson & Hammersley, 2007, p. 39.). Referring to this case study, if the researcher only observes one selective public space of the SSTEC, the result might be a bias in the analysis of people's behaviours, therefore, it needs to expand the scope of research locations to ensure observing types of people's behaviours as many as possible.

Being aware of the importance of context (place), regarding the participant observation, the researcher plans to observe in different public spaces within the eco-city by two standards of classification. On the one hand, by types of people's activities, including spaces with stationary activities (e.g. sitting and standing) and spaces with active activities (e.g. exercising and playing). On the other hand, it not only includes the crowded areas, such as city squares and community parks, where one can almost reflect conditions and objective laws of people's outdoor activities, but also includes the neglected areas, such as plaza space and street space within the eco-city. Moreover, for the street-intercept survey (questionnaire), in order to get more response rate, the relaxing place (e.g. specific public place with benches and with the shade where people can sit comfortably to do the questionnaire) is the first choice of the survey's location. Also, choosing an informal site with a relaxing atmosphere contributes to conducting the survey.

### d. **Evaluation Criteria Used in the Pilot Study**

An important purpose of the pilot study is to gain a broad understanding of the whole research site by researcher's observation, such as five elements (nodes, paths, districts, landmarks and edges) proposed by the Kevin Lynch, and to narrow down representative places in the next stage for data collection. An effective evaluation criterion is critical for the researcher to percept the place and to identify which are representative public spaces used for the further research.

The 12 quality criteria (*See* Fig. 4.3, *Appendix C and Appendix D)*, proposed by Jan Gehl, have been regarded as the guideline to create and to design public spaces in the urban context where people have higher needs about functional and desirable public space, such as comfort and enjoyment. These criteria, based upon Gehl's years of research, contribute to defining people's experiences of daily lives within the public realm and how urban design can help to enhance that experience. As *Gehl Architect Company* (2015) illustrated in its research report, 'an inviting space that encourages public life has elements of protection, comfort and enjoyment' (p. 11). Each of the category includes different criteria, and each of criteria has covered one aspect of various people's interaction and their experience within the public space.

### Protection (Safety)

To encourage users to utilize public spaces, protection against climate and environment is in vital. Three aspects of this evaluation have provided a well-rounded description about what good protection is, involving protection against traffic and accidents, protection against crime and violence, as well as protection against unpleasant sensory experiences.

### Comfort

'People are generally attracted to places they are comfortable being at, whether it is sitting, standing, walking, or talking to others, people want to be in an environment that makes them feel comfortable and welcomed' (Gehl, 2010, p. 176). Thus, for an environment to be welcoming, inviting and comfortable to everyone, it is of great importance to understand the need of differ user groups and to provide different usages for different purposes. The criteria in this element involve opportunities to walk, to stand and to stay, to sit and to see, to talk and to listen, as well as opportunities to play and exercise. Each criterion has its own focus point, and each is equally significant to providing comfort for people within the public space. In this category, different kinds of opportunities have been determined and measured by the amount and types of social infrastructures.

### Enjoyment

As Gehl (2010) stated, 'for an environment to be enjoyable, having human-scale elements provides people a sense of security without feeling lost and a sense of belonging compared to being surrounded by buildings or objects' (p. 178). The criteria in this element involve measurement at human scale, opportunities to enjoy the positive aspects of climate, as well as aesthetic qualities and positive sensory experience. These three criteria have outlined the basic design manual about what an enjoyable public space should have and provide to people.

## The 12 Quality criteria proposed by Jan Gehl

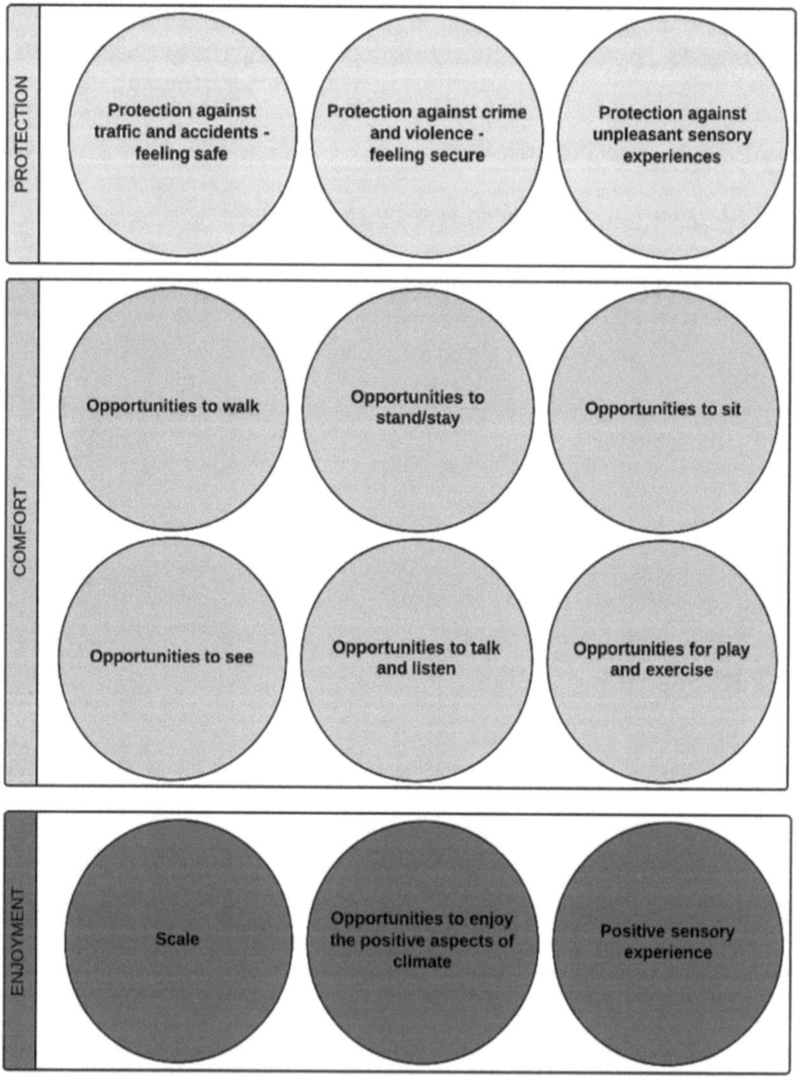


(Image is drawn by the researcher. 3 August 2020.)

(2) Eco-City Documents Analysis

Document analysis aims to better understand limits about the conceptualization of humanistic planning to be implemented in the SSTEC project. Also, it is functioned as the most important complement of in-depth interviews. These documents not only reflect how Chinese national government, Tianjin local government and relevant planning agencies implement the people-oriented policy instructions in their practical work, they also contain information on outcomes. Relevant documents are chosen by the following four dimensions, 'authenticity for ensuring the unquestionable authoritative source of evidence, credibility for looking at the producer of those documents and focuses on the accuracy, representativeness for ascertaining the typicality, and the meaningful documents with its evidence understandable and apparent' (Scott, 1990, p. 6).

Document analysis in eco-city planning research provides access to relevant policy documents, facilitates the identification of planning practices, enables evaluation of policy implementation, tracks progress and long-term impacts, supports comparative analysis and provides valuable insights for policy recommendations and future's planning efforts. It serves as an effective method for studying and advancing sustainable urban development practices within the context of eco-cities. Selected documents are formulated by different actors for different purposes. Identification of any commonalities, through the process of context analysis, does not in itself justify their being grouped together as a coherent target for document analysis. However, Fairclough's (1989a, 1989b) framework for analysing discourse offers the rationale for doing so—he regards texts, including spoken and written language and associated visual imagery as 'the form of action' (Fairclough, 1989a, 1989b, p. 9). Also, he regards discourse is one form of social practice shaped in turn by broader social structures, as he states, 'the whole is society, and language is one strand of the social' (Fairclough, 1989a, 1989b, p. 23).

## Advantages of Document Analysis Applied in This Research

| *Advantages* | *Detailed explanation* |
|---|---|
| Offers access to relevant policy documents | Relevant documents offer valuable sources of information on planning goals, strategies and implementation mechanism of eco-city initiatives. Such access provides comprehensive understanding of approaches used in eco-city planning |
| Facilitates the identification of planning practices | Researcher can identify best practices in eco-city planning by examining documents from successful eco-city projects worldwide, which include innovative approaches, effective strategies and lessons learned from various contexts. By identifying and studying these best practices, researchers can inform the implementation of eco-city plans in other locations |
| Enables evaluation of policy implementation | It enables researchers to evaluate effectiveness and outcomes of eco-city planning policies. By examining planning reports, progress assessments and performance indicators, researchers can assess the extent to which sustainability as well as people-oriented goals are being achieved. They can evaluate whether planning policies and measures implemented are producing desired environmental, social and economic outcomes. Such evaluation helps to identify both strengths and weaknesses of improvement in eco-city's initiatives |
| Tracks progress and long-term impacts | It allows researchers to track the progress of eco-city projects and assess their long-term impacts. By analysing annual reviews and monitoring data, researchers can evaluate the evolution of eco-city planning over time. Also, it allows researchers to measure the effectiveness of sustainable and humanistic measures, to assess environmental performance, and to identify trends or patterns in the implementation of eco-city initiatives |

(continued)

(continued)

| *Advantages* | *Detailed explanation* |
|---|---|
| Supports comparative analysis | It allows researchers to identify similarities and differences in planning approaches, and outcomes. Comparative analysis helps to understand factors that contribute to successful eco-city planning and contextual factors that influence outcomes |
| Provides valuable insights for policy recommendations and future's planning efforts | By identifying gaps, challenges and influencing factors in existing policy documents, researchers can propose appropriate improvements. Findings from document analysis can provide evidence-based recommendations for enhancing the combination of sustainable and humanistic aspects of eco-cities |

(Source: Information is summarized by the researcher.)

As discussed in Chapter 1, one defining feature of the eco-city phenomenon is its globalized nature, characterized by its geographical spread, the increasing involvement of international practitioners and policy-makers, and also the claimed global relevance of individual initiatives for mankind's future. While limiting the investigation to any particular country or region would therefore compromise its value, the variety of policy approaches encompassed by the eco-city also prompts the inclusion of a wide range of initiatives. Relative documents, being formulated by various types of actors for different purposes, are deemed as the worthwhile object of study, precisely because it is institutionally ratified. The guiding principle behind the selection method aims to facilitate the inclusion of a wide variety of practices, policies and contexts. Also, discursive framings and concrete prescriptions of these documents play significant roles in shaping practical Chinese eco-city development. Last but not least, search protocol focuses on planning goals and mainstream initiatives of Chinese eco-city planning and design, which might be expected to reflect and to facilitate a greater degree of practical influences on Chinese green urbanization and eco-city development.

The precise documentation consulted in case study appears in Table 4. 5. As the table shows, the nature and status of these documents necessarily varied, given their divergent functions and scalar coverage. More positively, the inclusion of different types of documentation would actively contribute to the overall goal of capturing the variety of eco-city conceptualization.

Documents have been analysed as the complement of qualitative interviews and observations. These documents not only reflect how Chinese local governments implement the top-down eco-city policy instruction in their practical work according to the local context of eco-city development, but also contain information on roles, approaches and outcomes of humanistic planning for Chinese eco-city development. In addition, relevant documents have been chosen by following the guidance of four dimensions according to Scott (1990), 'authenticity' for ensuring the 'unquestionable authoritative' source of the evidence, 'credibility' for looking at the producer of those documents and focuses on the 'accuracy', 'representativeness' for ascertaining the 'typicality' as well as the 'a typicality' associated with the data, and the 'meaningful documents' with its evidence understandable and apparent (p. 6). These documents include five categories:

–Global academic sources (journals and reports on eco-city theme).
–Plans, including municipal and district level master plans, land-use plans from district level (Tianjin Binhai New District).
–Local news in recent five years with the focus on public events and officials' public speeches on the development process of eco-city through the internet.
–Government working records, such as working diaries, summaries and notes, officials' speeches and other related documents.
–Planning policies (esp. regarding indicator system), regulations and laws.

These documents have been accessed in three ways:

–Some of them are accessed through the official government website, which is available to the public.
–Some of them are accessed through key informants who participated in the redevelopment project in the case.
–Some of them are accessed through the *SSTEC Construction Bureau* and *SSTEC Administrative Committee*, such as detailed land plans and working diaries, for which access needed was to be approved by the officials.

Details on search protocol and how to select documents and journal articles are shown in the following table. And detailed operation methods revolve around the core word *eco-city development*.

**Research Protocol and Methods for Document Section**

| *Steps* | *Specific operation methods* |
|---|---|
| Define the research focus | As eco-city development is a broad topic, it is crucial to define subtopics. At the initial stage, specific aspects are divided into several categories, including urban planning approach, green infrastructure design, social equity consideration, economic viability and technologies innovation |
| Identify keywords | Create a list of keywords or key phrases related to research focus, which has encompassed various aspects of eco-city development. For instance, relevant keywords include *eco-city, sustainable urban development, green urbanization, urban sustainability, smart cities, circular economy, people-centred planning* and so on |
| Select relevant database | Consider database that cover disciplines such as urban planning, sustainability, environmental psychology. Firstly, one example of useful database includes Web of Science, Scopus, Google Scholar, Urban Studies Abstracts, Environmental Impact Statement. Secondly, Chinese government websites at the national, provincial and municipal levels often publish reports, policies and guidelines related to eco-city development. These websites are valuable sources of official documents and policies that highlight the government's strategies and initiatives. Thirdly, to stay updated with daily news on eco-city development, it is necessary to explore popular Chinese news platforms such as SSTEC WeChat Official Account, China Eco-City Academy Website. These platforms often cover the latest development, projects and policies related to sustainable urbanization in China |
| Review literatures | At this stage, going through abstracts and full-text documents determine their relevance to main research questions that are pertinent to this research. Taking notes for each source and highlighting key points to organize and synthesize the information gathered |

(Source: Information is summarized by the researcher.)

The first two categories mentioned above have been collected and reviewed before the fieldwork, they are together provided an overall image to understand the rationale and political tasks of the SSTEC's planning and (re)development programmes from different aspects, such as geographical and political conditions, project vision and planning ambitions, which will be further discussed in the next chapter. The last category, used to understand social and political context of this eco-city project, is acquired during online interviews. These documents directly displayed how local government and its agencies implement policies related to the humanistic planning concept. They are used to find the breakthrough about obstacles that limit the SSTEC to meet the goal's development about considering people's needs in the actual planning and design process, and to explore valuable insights about potential influences of people-place relationship.

### Four Categories of Documents Related to the SSTEC

| *Category* | *Main documents* |
|---|---|
| Global academic sources (journals and reports) | Flynn, A. and Yu. L. et al. (2016). Eco-Cities, Governance and Sustainable Lifestyles: The Case of the Sino-Singapore Tianjin Eco-City. *Habitat International*. 53: 78–86<br>Catherine, C. I-C. et al. (2016). A Green Leap Forward? Eco-State Restructuring and the Tianjin-Binhai Eco-City Model. *Regional Studies*. 50(6): 929–943<br>Caprotti, F. (2014). Critical Research on Eco-City? A Walk through the Sino-Singapore Tianjin Eco-City, China. *Cities*. 36(2):10–17<br>Caprotti, F. et al. (2015). Eco for Whom? Envisioning Eco-Urbanism in the Sino-Singapore Tianjin Eco-City, China. *International Journal of Urban and Regional Research*. 39(3): 495–517<br>World Bank (2009). *Sino-Singapore Tianjin Eco-City: A Case Study of an Emerging Eco-City in China (Technical Assistance Report)*. [Online.] Available at: https://openknowledge.worldbank.org/handle/10986/28143. [Accessed on 17th February 2019)<br>World Bank. (2010). *Eco2 Cities Brochure*. [*Online*.] Available at: http://siteresources.worldbank.org/INTEASTASIAPACIFIC/Resources/Eco2_Cities_Brochure.pdf<br>Sino-Singapore Tianjin Eco-City Administrative Committee, China Eco-City Academy and Centre for Liveable Cities. (2021). A Partnership for a Smart and Sustainable Future: The Sino-Singapore Tianjin Eco-City. [Online]. Available at: https://www.clc.gov.sg//docs/default-source/books/a-partnership-for-a-smart-and-sustainable-future-the-china-singapore-tianjin-eco-city.pdf |
| Local news (three typical online news) | The Zhongxin Avenue will be widened. 11th July 2018. [Online]. Available at: https://www.sohu.com/a/240637196_737811<br>The SSTEC has formulated the action plan about the upgraded index system (2.0 version). 26th September 2020. [Online]. Available at: https://www.sohu.com/a/421039528_620823<br>Eco-City needs to be based around communities, not technology. 21st October 2013. [Online]. Available at: https://chinadialogue.net/en/cities/6427-eco-cities-need-to-be-based-around-communities-not-technology<br>'New KPIs' to lead 'New Journey': The Upgraded Version 2.0 KPIs was approved. 25th September 2021. [Online]. Available at: http://www.tianjinecocity.com/News/publications_show_detail/795?lang = english |

(continued)

(continued)

| *Category* | *Main documents* |
|---|---|
| Masterplan | Masterplan of the Sino-Singapore Tianjin Eco-City (2008–2020 Version) (formulated by *China Academy of Urban Planning Design*, *Tianjin Urban Planning Design Institute*, *Urban Redevelopment Authority of Singapore*) |
| Planning policies | Development Retrospect of the Sino-Singapore Tianjin Eco-City (formulated by *China Eco-City Academy*)<br>Reply on the 2.0 Version of Indicator System<br>Report of the Implementation of Indicator System of the Sino-Singapore Tianjin Eco-City (2020) |

(Source: Information is summarized by the researcher.)

The process of *Chinese New-Type Urbanization* and the implementation of ecological development policies, as Chapter 1 discussed, have jointly promoted the eco-city development. With the continuous development of urban construction and operation at different stages, users' needs and demands are constantly changing. Being faced by users' actual needs and problems encountered in the operation process, the eco-city shall be continually improved and updated, also functions and services need to meet users' demands accordingly.

After organizing information into categories related to the evaluation dimension of users' demand, this research regards Chinese national standard *Assessment Standard for Green Eco-District (GB/T51255-2017)* as the national masterplan for guiding, assessing and certifying humanistic sustainability as well as environmental performance of Chinese eco-districts in the following four ways, including comprehensive assessment criteria, performance measurement, consideration of social and economic aspects, adaptability and scalability. Firstly, this standard includes a set of criteria that has covered various aspects of sustainability and environment performance. These criteria are developed to address specific needs and demands of eco-district stakeholders which include developers, urban planners, government agencies and residents. Secondly, by providing clear guidelines and metrics, this standard outlines a methodology for assessing and measuring the performance of eco-districts. It also enables users to evaluate their projects' sustainability and environmental impact objectively. Through this method, it allows stakeholders to understand how their projects align with desired objectives and meet the demand of a sustainable eco-district. Thirdly, this standard not only has addressed environmental aspects but also considers social and economic dimensions of eco-districts, which reflect the demand for a holistic approach to sustainability by considering factors like social equity, economic viability and job creation. By encompassing these dimensions, this standard recognizes the importance of balancing environmental concerns with social and economic development. The last point highlights the need for humanization in standard setting as this standard accounts for diverse contexts and scales of eco-district projects by offering flexibility for users to adapt the criteria and indicators to their specific circumstances while maintaining the overall integrity of assessment framework. That is, this approach has recognized unique demands and needs for different projects, allowing users to implement sustainable practices effectively regardless of project's size or location.

Since the Chinese national standard *Assessment Standard for Green Eco-District (GB/T51255-2017)* has been promulgated and implemented, the standard has strong authority, a large number of credits and rich evaluation content. To evaluate in what ways green eco-district planning takes the humanistic aspect into the overall consideration rather than focusing much on technologies, the user demands library, therefore, has been proposed according to the national standard by doing document analysis; at this stage, skimming (superficial examination), reading (thorough examination) and interpretation.

Regarding the national standard *Assessment Standard for Green Eco-District (GB/T51255-2017)*, from the perspective of evaluation dimension, it mainly includes dimension about government management and dimension about user demand. For instance, the credit *4.1.1 urban planning should meet the urban and rural planning requirements of the region* can be classified as the content of government management dimension, while *4.2.5 public service facilities in residential areas have better convenience* can be classified as the content of user demand dimension. For document analysis and evaluation, this research starts from the perspective of user dimension by combining and categorizing the content of national standard *Assessment Standard for Green Eco-District (GB/T51255-2017)*, and initially categorizes the user demand library into six major areas which involves land use, ecological environment, green building, energy utilization, green transportation and humanities.

## Dimension of User Demand Library in Eco-City

| *Serial no* | *Major categories* | *User demand item* | *Content overview* |
|---|---|---|---|
| 1 | **Land use** | Convenient public service facilities | Kindergartens, nurseries, primary and secondary schools, pension service facilities, health service centres, commercial service facilities and other facilities around the residential area, which are convenient and accessible |
| 2 | | Opening of public space | Public open space is set up in urban area for residents' activities |
| 3 | | Green space opening | The ecological land and urban green space with a certain scale and reasonable layout shall be maintained in the urban area |
| 4 | **Ecological environment** | High greening rate | Three-dimensional greening is implemented in the maintenance and management of all kinds of garden green space are good, and the greening coverage rate is high in the urban area |
| 5 | | The rain water quick absorption | Implementation of sponge city construction, rainstorm without urban waterlogging |
| 6 | | Avoid soil pollution | There is no soil pollution or the soil pollution control is completed and up to standard |
| 7 | | High quality of domestic water | The quality of municipal domestic water reaches the national standard |
| 8 | | Good outdoor air quality | Annual excellent days of air quality reached more than 240 days, average days of reaching standard of PM2.5 reached more than 200 days |
| 9 | | Good outdoor sound environment | The outdoor sound environmental quality of the area meets the national standard |

(continued)

(continued)

| *Serial no* | *Major categories* | *User demand item* | *Content overview* |
|---|---|---|---|
| 10 | | Refuse classification | Classified collection, closed transportation and effective treatment of garbage |
| 11 | **Green building** | Newly built green building | Choose to buy green residential buildings or work in green office buildings |
| 12 | | Green renovation of existing buildings | The residential buildings or office buildings that have been renovated to meet the requirements of national standard for green buildings |
| 13 | | Green construction | The projects under construction meet the requirements of green construction and reduce the pollution and impact on the surrounding environment |
| 14 | | Green operation | Green operation is adopted for residential buildings or office buildings, which can reduce operating costs and provide good building environment and high-quality services |
| 15 | **Energy utilization** | Energy measurement | Charging for heating (cooling) consumption |
| 16 | | Reclaimed water use | Residential or office buildings have the conditions to utilize reclaimed water |
| 17 | **Green transportation** | Public transport system | With high coverage of bus stops, bus lanes are set up, and the public transport system has humanized service facilities (such as guidance, barrier-free access, sunshade, seats, etc.) |
| 18 | | Bicycle transportation system | Urban bicycle lanes are continuous, with reasonable width to form a shady road, and complete road supporting facilities (such as guide signs, safety, rest, sanitation and other facilities) |

(continued)

(continued)

| *Serial no* | *Major categories* | *User demand item* | *Content overview* |
|---|---|---|---|
| 19 | | Walking system | The walking system is continuous, combined with surrounding functions, environment, landscape and public space, with complete supporting facilities (such as lighting, guiding signs, safety, rest, sanitation and other facilities) |
| 20 | | Charging facilities for new energy vehicles | Electric vehicle charging facilities should be built in residential buildings, and public parking lots |
| 21 | | Parking spaces | The underground parking lot and public parking lot are provided with sufficient parking spaces, and the parking resources are efficiently utilized, and the parking information is timely and unobstructed |
| 22 | **Humanity** | Elderly service facilities | The community service network for the aged is completed and there are enough beds for the aged |
| 23 | | Encourage green lifestyle | Formulate preferential measures to encourage residents to purchase energy-saving and water-saving product |

(Source: Information is from Chinese national standard *Assessment Standard for Green Eco-District (GB/T51255-2017)*

### *The Second Layer: The Multiple-Method Qualitative Data Collection Approach*

An explanatory case study, being used to 'do causal investigation and to understand the complex social phenomena' (Yin, 2014, p. 6), is designed to apply in the second-layer research strategy. It is the most important stage of this research, via the multiple-method data collection to explore the people-place relationship in the real-world eco-city project. It is worth mentioning that typical qualitative research methods including participant observation and questionnaires at this stage respond to *Objective Three* and *Research Question Two*, by exploring the public life in public spaces from the perspective of people's experience to provide the foundation of preconceptions for further data collection.

The second-layer stage, using the *Public Space and Public Life (PSPL) Survey* proposed by Jan Gehl—the approach utilizing empirical analysis to understand how the built environment can promote well-being and effect social interactions between people. Specially, the *PSPL Survey* adopts further on-site observation and qualitative questionnaires, attempting to verify data collected from the pilot study, as research data of the first layer tends to be occupied by the subjective perception of researcher's feelings and judgement, such as the evaluation about spatial quality of several public spaces within the SSTEC.

On the one hand, this stage of data collection focuses on the public space analysis by means of mapping and on-site counting, aiming to explore types of people's outdoor activity and users of public spaces (e.g. different ages and occupations). In addition, it emphasizes on the public life survey by means of on-site observation, aiming to explore a variety of spatial elements that can potentially stimulate or prevent people to spend more time in public spaces. On the other hand, the street-intercept survey by the qualitative questionnaire, being functioned as the supplement to observation at data collection stage, aims to explore people's more detailed perspectives and objective descriptions about their perception and attitudes (defined as 'invisible' needs) towards public places, as well as their comments on social infrastructures within the eco-city. That is, to investigate and to identify criteria and causes of good or bad spatial qualities in their eyes.

### *The Third Layer: Data Analysis and Further Data Collection*

The last research stage of a series of in-depth semi-structured interviews with planning practitioners and management operators of the SSTEC project is based upon the analysis of all research data from the former two layers. It is concerned about answering the last main research question—understanding whether and how does the case study project consider impacts or implications on the interactive pattern between people and place. At the end of this stage, by combining all findings, the overall understanding of humanistic planning under the context of ecological development aims to be achieved, and potential recommendations on people-oriented planning principle and procedure is devised. An explanatory case study, being used to 'do causal investigation and to understand the complex social phenomena' (Yin, 2014, p. 6), is designed to apply in the second-layer research strategy. It is worth mentioning that typical qualitative research methods including participant observation and in-depth interviews at this stage respond to *Objective Three* and *Research Question Two*, by synthesizing key findings to explore the existing obstacles that may limit Chinese eco-city project as well as the existing planning initiatives to meet the development goal about considering people's needs.

## Research Methods in an Interdisciplinary Context: Public Space and Public Life Survey (PSPL) and In-Depth Interviews

In the early stage of the development of *Environmental Psychology* discipline, researchers often use methods and tools borrowed from the experimental psychology to conduct research. For example, it is the first time for William H. Whyte to use the observational method in the natural situation—a three-year study on the use of public space in the city centre of New York City. Research methods of *Environmental Psychology* adopt some content that is much related to disciplines of sociology and psychology, with particular attention to the interaction between people and the physical environment. It mainly focuses on the collection and analysis of spatial information; for instance, Bechtel et al. (1987) summarized 13 specific methods about data collection, such as open-ended or structured interviews, cognitive or behavioural maps, participatory observation, questionnaire survey, psychological test, adjective check, etc. Based on the setting of two main research questions, this research has selected several methods for SSTEC's field work and for further data analysis.

The *Public Space and Public Life (PSPL) Survey*, a tool to document the existing people's behaviours, helps to understand the types of people's activities, how people use the public space and barriers that may inhibit socializing in the public space (Zhao et al., 2012, pp. 34–35). It creates an opportunity for the researcher to collect humanistic data and then to make people's needs visible. Also, the *Public Space and Public Life Survey*

*(PSPL) Survey* considers how existing people's behaviours can inform the design process to make a place more liveable, walkable and inviting to all. As part of the survey, pedestrian count and stationary activity surveys are used to examine detailed information about what people do and where they do. By using this technique, the researcher also assesses the condition of outdoor seating and other factors that affect people's satisfactions/ dissatisfactions towards the public space.

### (1) **Rationale for Choosing This Research Method**

Participant observation is useful as a method of 'acquainting designer and researcher with the lifestyle of people for whom a design is intended' (Bechtel et al., 1987, p. 19). In this research, participant observation is functioned as the fundamental research tool to collect data about spatial elements of the eco-city and people's outdoor activities within the public space. It helps to understand what kind of activities people engage in, how they use the public space in their neighbourhoods, as well as a variety of spatial elements that can potentially stimulate or prevent people to spend more time in public spaces. In addition, data collection by means of participant observation aims to explore in what ways and to what extent can public spaces become more connected, liveable, walkable and also be available to people that make residents enjoy outdoor activities in eco-city.

Observational data are summarized graphically, and complemented with questionnaire information provided by interviewees (esp. residents living in the eco-city), along with photographs, to provide a full picture of each selected research location within the SSTEC. Although adopting various types of data and methods contributes to providing a degree of valuable 'methodological triangulation' (Guion, 2002), it is acknowledged that analysis requires a degree of subjective judgement on the researcher's perspective (raising questions about its replicability). The approach, combined researcher's observation and eco-city's residents' feedback, would have been improved through 'investigator triangulation' with validity more firmly established through the concordance of findings.

### (2) Limitations

The main limitation of participant observation in this case study tends to be the influence of being observed people, as a result not all activities can be observed. To reduce this potential impact, on the one hand, the researcher intends to start the observation in a covert role to easily gain access to do mapping and on-site counting. Starting with a covert role contributes to easily observing people's outdoor activities, as 'they are less likely to adjust their behaviours because of the researcher's presence' (Bryman, 2016, p. 406). After mapping and on-site counting, the researcher plans to move from covert role to an overt one for participating in some outdoor activities with residents. For instance, the researcher intends to join the square-dancing[1] activities group to be better integrated into their everyday lives and to prepare for distributing questionnaires. Having their consents during conducting informal discussions (causal talking) with some people when taking part in activities is essential followed by citing the data elicited from their talks. On the other hand, openness is of great importance to communicate with people when doing the covert observation—if someone asks about why observe them, the researcher should attempt to explain this research project in brief.

## *Qualitative Questionnaire (Street-Intercept Survey)*

To some extent, the self-completion questionnaire and the structured interview are similar to social research. The obvious difference between them is that, with the self-completion questionnaire, respondents should read and answer each question by themselves as there is no interviewer to ask questions (Bryman, 2016, p. 233), making research instrument easy to follow.

### (1) Rationale for Choosing This Research Method

[1] In China, square dancing refers to an exercise routine performed to music in squares, plazas or parks. It is popular with middle-aged and retired women who have been referred to as *dancing grannies*.

The street-intercept survey by qualitative questionnaire, as the supplement to participant observation (mapping, counting and on-site observation), contributes to obtaining people's more detailed perspectives and objective descriptions of those 'invisible' needs towards public places and social infrastructures within these places.

(2) **Limitations**

Time consumed to collect data and low response rate might be two main limitations of this research method, as it takes long time to get enough response rate and reliable information. To tackle these potential issues, since the research object only focuses on the group of residents living in the eco-city, snow sampling has been considered to be used for collecting data, as it to some extent can enlarge the sample size and can improve the effective feedback rate. Also, web-vision questionnaire will be used to increase the response rate if the street-intercept one gets low-rate response. Moreover, time consumed to analyse data might be the potential limitation of this qualitative research method, as the researcher has to read answers and try to put them into categories by coding. To overcome this issue, more closed questions by ticking answers and less open-ended questions are applied when formulating the questionnaire.

### *In-Depth Semi-Structured Interviews*

Interviewing, 'a conversation with a purpose' (Kvale & Brinkmann, 2009), has been deemed as the overall strategy as well as the most common and effective research approach to gather first-hand data in a case study. Three types of interviews are divided as structured, semi-structured and unstructured interviews, based upon the extent of control that interviewer has over the interaction.

Specifically, unstructured interview does not reflect any 'preconceived theories or ideas' (May, 1991, p. 188), that is, the interviewer starts with an opening question and continues the conversation on the basis of interviewee's response. In this sense, it is usually time-consuming and difficult to participate in, as 'the lack of predetermined interview questions provides little guidance on what to talk about' (Gill et al., 2008, p. 291). In contrast, structured interview is 'verbally administered questionnaires, in which a list of predetermined questions will be asked, with little or no variation and with no scope for follow-up questions to response' (*ibid.*) However, this type of interview only allows for limited participant responses, being of little use if depth is required.

Semi-structured interview has been seen as a more balanced approach—a wide range of information can be effectively collected during interviewing, but more time-saving compared to the unstructured interview, it also can make the interviewer get access to the focus of attention, as the structured interview does. In the preparation stage of semi-structured interview, the outline consisting of a clear set of topics and open-ended questions should be well prepared in advance. And during the interviewing, the interviewer can make moderations to those prepared questions and ask more detailed ones according to the interviewee's responses, making the interviewee express his/her own opinions freely.

### (1) **Rationale for Choosing This Research Method**

Semi-structured interview has been chosen to understand whether and how does the case study project consider impacts on the interactive pattern between people and place, by interviewing planning practitioners and management operators of the SSTEC project, and also to explore constraints and planning initiatives on effectively promoting the people-oriented planning and design concept.

### (2) **Skills Used for Interviewing**

Case study requires an inquiring mind during data collection, not simply before or after the activity. The ability to pose good questions is therefore a prerequisite for case study researchers, thus, the design of interview questions has to be carefully taken into consideration. The *Interview Topic Guides* (see *Appendix G*) mentions several topical areas, and each of these is introduced by an open question and ends with a confrontational question. Also, some so-called 'theory-driven questions', which are oriented to the literature about related topic or based on the researcher's theoretical presuppositions, aim to be asked to verify previous research findings from observation and questionnaires.

Considerations should be taken into account for each interviewing stage. Firstly, in order to be objective, at the beginning of each interview, the researcher aims not to ask any specialized academic knowledge about the people-oriented planning concept (relationship between people and place), but rather to ask how planning and design methods related to this interactive pattern are used during the post-operational stage of the SSTEC project. In addition, since the researcher has finished the observation, in order to avoid taking things for granted, posing open questions and keeping an open mind about responses are of great importance. Secondly, during the interviewing process, 'follow-up questions are often posed, which can make the researcher obtain more inside information' (Kvale & Brinkmann, 2009, p. 36). Therefore, the process of active listening requires the interviewer to listen to what has been said by the interviewee and how it is said. Thirdly, after finishing, each interview plans to be rounded off by a debriefing, asking about whether the interviewee has more information to say, which offers an additional opportunity to deal with issues he or she has been thinking or worrying about during the interview.

To understand an interviewee correctly in a high-context culture is not an easy job for an outsider (Hall, 1976, p. 108). In a high-context culture, many things are left unsaid, letting the culture to explain. A few words can communicate a complex message very effectively to an in-group, but less effectively to those outside that group. While in a low-context culture, the communicator needs to be much more explicit and the value of a single word is less important. Since the researcher fully understands the Chinese high-context culture, listening carefully in order to capture what the interviewee meant in that situation, rather than simply noting down their exact words.

## Thematic Analysis

This research has adopted thematic analysis as an approach to analyse and to interpret themes of data. The reason for choosing thematic analysis as it not only summarizes and reorganizes interview transcripts, more importantly, it contributes to connecting with research objectives and questions to identify and to interpret key features of research data.

Related planning practitioners were asked to contribute their knowledge and viewpoints on the SSTEC project in a colloquial way. However, information about their practical experience has not been linked logically, thus, it is necessary to establish a framework for analysis. During the process of data analysis, interview transcripts are all reorganized in order to 'move from discussion of issues within one area to discussion of issues within another, with each area logically connected to others' (Weiss, 1995, p. 154).

### (1) Transcription and Translation

All interviews are audio-recorded after which the first step of data analysis is to transcribe all information into written form. Instead of what has been said by interviewees, 'meanings of utterances are profoundly shaped by how they said' (Sarangi and Robert, 1999). Therefore, the researcher interviewed with the special intention of features of interviews, such as speed and tone of voice, as well as body language and facial expressions, which are all crucial for interpreting. All above-mentioned information is coded together with the transcript of interviews, and the final transcript has been scrutinized several times before analysis in order to make the researcher be familiar with the data.

All interviews with planning practitioners in this research are conducted in Chinese. After transcribing into the written form, they are translated into English. Thus, linguistic issues are inevitable during this process—some Chinese metaphorical expressions could not be directly translated into English, consequently, part of interviewees' original meanings might be lost in the compromise of using similar English vocabularies. To reduce the likelihood of such linguistic issue, detailed interpretations by considering features of interviews have been provided where meanings could not be clearly expressed in English.

### (2) Coding

Various understanding of the coding exists in different researches, but in common, it is a process of describing 'relations of materials collected from the research, labelling and categorising data in analysis' (Boyatzis, 1998, p. 42). Thematic coding, a form of qualitative analysis, refers to recording information which has been linked by common themes and establishing 'a framework of thematic ideas' (Gibbs, 2007, p. 39). It is adopted to establish a framework in which codes are grouped into different research themes. In Chapters 1 and 2, research aim, objectives and conceptual framework have set out both key areas and border themes required to be researched. Also, interviews are conducted in a semi-structured way and guided by the list of research questions. Detailed coding process of this research is shown in the following table. And coding development continuously proceeds until no new codes are being generated.

**The Coding Process**

1. Browsing through all English transcripts as a whole and making notes about the border topics interviewees talked about
2. Reading all transcripts to generate codes from simple descriptions to conceptual categories. Highlighting relevant key words, noting similarities or differences from the previous research about literature reviews, and putting more emphasis on interviewees' viewpoints they repeated several times
3. Putting all codes into different research themes
4. Making the comparison between/within codes
5. Producing the code list by distinguishing important codes

(Source: Table is drawn by the researcher.)

### (3) Thematic Development

After generating some significant codes, the code framework has been developed, however, it lacks explanation about how the coding links with the broader ideas of this research; it is thus followed by a thematic development. Specifically, during the thematic analysis process, inductive and deductive thematic development methods are all used to interpret research data—the theory-driven deductive approach generates themes referring to the central content and interpretive meaning of ***humanistic planning under the idea of ecology*** concept from literature reviews, while the data-driven inductive approach helps themes to be generated from analysis process of the selected case study.

## Potential Ethical Issues and Solutions

### *Informed Consent*

Research ethics have been seen as the main consideration when doing fieldwork outside the UK, and informed consent is one of the most essential concerns. As Marshall (1992) indicated, 'the nature and process of obtaining informed consent in non-western cultures can be problematic. Regardless of the specific cultural setting, the implementation of informed consent may become unethical in local contexts' (p.2). Specifically, in China, asking people to sign the agreement seems impolite as Chinese people attach importance to relationship (*guanxi* in Chinese). For them, only big issues need the writing agreement, and if the researcher forces them to sign the writing informed consent, situation of wounding people's feelings may occur. Thus, with the acknowledgement of cultural characteristic, the researcher attempts to use a different tone or different linguistic terms to different people in order to show respect.

In terms of observation, if someone asks about why observe them, the researcher will attempt to inform them orally about the purpose and main issues of this research rather than specific details. The reason for not informing them everything is due to the difficulty about 'pre-defining the exact nature and scope of study' (Marshall, 1992, p. 2). In terms of interview, each interviewee will be given a brief introduction about the research purpose at the beginning of interviewing. When finishing the interview, a request will be made to sign the formal written informed consent in Chinese, followed by sending the finished interviewing transcript to each interviewee by email.

### *Covert Observation*

The most obvious ethical issue of this research has been seen as the covert observation, due to 'its potential harm to the subjects, damage to the general reputation of sociology, and the possibility that it might close further avenues for research' (Bulmer, 1982, p. 18). However, covert research for observing people's outdoor activities in public spaces within the eco-city is the revealing and necessary approach if it conducts in an appropriate way. As Bulmer (1982) indicated, 'covert participant is not a widely-used research technique, but its use clearly brings out some of the ethical principles that guide the conduct of social research' (pp. 23–24).

The rationale of applying covert observation into this research is shown as below. First and foremost, the priority of sociology defines as exploring the unseen and searching for the truth, and the social researcher is entitled to adopt covert methods in his/her research if the social scientific knowledge about society outweighs the potential risk during the fieldwork. In addition, 'any method that moves us towards that goal, without unnecessary harm to subjects, is justifiable' (Bulmer, 1982, p. 25). That is, the sociologist has the right to make observations about anyone in any setting to the extent that he/she bears the scientific intents and purposes in mind. In this research, covert observation can access the real information about residents' daily lives, finding out what do residents do in public spaces within the eco-city.

### *Limitations and Possible Solutions*

*Limitation of Case Study Methodology*

The most common interrelated criticisms to the case study are seen as methodological rigour, subjectivity and external validity. This section summarizes above-mentioned concerns particularly related to the SSTEC case study, and provides solutions to dispel such doubts.

First of all, lack of preciseness is mainly due to the 'relative absence of methodological guideline' (Yin, 2009, p. 45). Consequently, increasing number of social science researchers have sought to develop their methodological techniques and epistemological grounding (Bennett & Elman, 2007). This research has adopted the multi-layer methodological framework to guide the whole process of data collection, and to increase the research rigour. Secondly, even though Hamel et al. (1993) criticized, 'the case study lacks of rigor in collection, construction and analysis of the empirical materials, ... such rigor is linked to the problem of bias ... introduced by the subjectivity of the researcher and others involved in the case' (p. 23). To reduce such bias, the researcher has attended training courses about interviewing skills before the fieldwork (detailed information is shown in the previous section named *Skills Used for Interviewing*), and has enough background knowledge of the selected case study by reading a large number of relevant documents. Last but not least, external validity emphasizes how applicable findings are to the real world, namely the generalization of research result. Flyvbjerg (2006) pointed out one of misunderstanding of the case study, 'one cannot generalise on the basis of a single case is usually considered to be devastating to the case study as a scientific method' (p. 224). Yin (2009) refuted that misunderstanding by illustrating the well-constructed explanation about differences between analytic generalization and statistical generalization—the previous one 'previously developed theory is used a template against which to compare the empirical results of the case study' (p. 38). From this perspective, as this research aims to explore the *conceptual-practical gap* in knowledge, research findings could be the reference for further studies and can replicate experience for other Chinese eco-city projects, even other types of cities adopting the people-oriented planning goal.

*Limitations at the Practical Level*

The main challenge of this case study design is related to the feasibility of interviews in practice. Specifically, on the one hand, some in-depth interviews were conducted online due to travel restrictions to China caused by the COVID-19 pandemic, in order to ensure enough valuable information to be gained as much as possible, the researcher has emailed the *Interview Topic Guides* (*See Appendix G*) to all relevant interviewees during communication stage before the actual interviews conduct online. Moreover, to ensure the accuracy of information, before coding, all interview transcripts (in Chinese written form) have been emailed to interviewees for double-checking. However, to compensate for this limitation, during the supplementary fieldtrip in June 2023, the researcher conducted additional face-to-face interviews. On the other hand, due to confidential reasons, some governmental officials were unwilling to release some information because of the sensitivity of their positions, such as oversensitive information about weakness of Chinese planning and design problems. With regard to this challenge, the researcher attempts to search for more reliable resources like media interview or statements made by other relevant officials on the same issue.

# Bibliography

Abada, T., Hou, F., & Ram, B. (2007). Racially Mixed Neighbourhood, Perceived Neighbourhood Social Cohesion, and Adolescent Health in Canada. *Social Science and Medicine, 65*, 2004–2017.

Adam, D., & Watkin, C. (2014). *The Value of Planning*. RTPI Research Report. Retrieved April 23, 2022, from https://www.rtpi.org.uk/media/1548/value-of-planning-full-report-2014.pdf

Aikenhead, G. S. (1997). A Framework for Reflecting on Assessment and Evaluation. *Globalization of Science Education: International Conference on Science Education* (pp. 195–199). Seoul, Korea the Korean Education Development Institute.

Alexander, C. (1977). *A Pattern Language* (1st ed.). Oxford University Press.

Alexander, C., et al. (1987). *A New Theory of Urban Design* (1st ed.). Oxford University Press.

Alfonzo, M. A. (2005). To Walk or Not to Walk? The Hierarchy of Walking Needs. *Environment and Behaviour, 37*(6), 808–836.

Altman, I., & Zube, E. H. (1989). *Public Places and Spaces* (1st ed.). Plenium Press.

Asian Development Bank. (2010). *Eco-City Development: New and Sustainable Way Forward?*. Retrieved April 28, 2022, from https://www.adb.org/sites/default/files/publication/27868/urbandev-prc-nov2010-ecocity.pdf

Aspinall, P., Mavros, P., Coyne, R., & Roe, J. (2015). The Urban Brain: Analysing Outdoor Physical Activity with Mobile EEG. *British Journal of Sports Medicine, 49*(4), 272–276.



Y. Wu, *Possibilities and Limitations of Chinese Eco-City Development*, New Perspectives on Chinese Politics and Society,
https://doi.org/10.1007/978-981-95-1958-3

Atkinson, P., & Hammersley, M. (2007). *Ethnography: Principles in Practice* (3rd ed.). Routledge.

Baker, R. A. (1968). The Future of Psychological Instrumentation. *Behavioural Sciences, 13*(1), 1–17.

Barton, H., & Kleiner, D. (2000). *Innovative Eco-Neighbourhood Projects.* Cited in: Barton, H. (2000). *Sustainable Communities: The Potential for Eco-Neighbourhoods* (1st ed., pp. 66–85). Earthscan.

Barton, S., & Pineo, R. (2009). *Human Benefits of Green Spaces. Cited. In: Barton, S. and Pineo, R. (2009). Sustainable Landscapes Series* (1st ed.). University of Delaware.

Beatley, T. (2000). *Green Urbanism: Learning from European Cities* (1st ed.). Island Press.

Bechtel, R., & Churchman, A. (2002a). *Handbook of Environmental Psychology* (1st ed.). Wiley & Sons.

Behrens, R., & Watson, V. (1996). *Making Urban Places: Principles and Guidelines for Layout Planning* (1st ed.). UCT Press.

Bentley, I., et al. (1985). *Responsive Environment XE "Responsive Environment" s: A Manual for Designers* (1st ed.). The Architectural Press.

Bennett, A., & Elman, C. (2007). Case Study Methods in the International Relations Subfield. *Comparative Political Studies, 40*(2), 170–195.

Bechtel, R. B., et al. (1987). *Methods in Environmental and Behavioural Research* (1st ed.). Van Nostrand Reinhold Company Inc.

Beck, U. (1992). *Risk Society: Towards a New Modernity* (1st ed.). Sage Publications.

Bell, A. P., Greene, T. C., Fisher, J. D., & Baum, A. (1984). *Environmental Psychology* (1st ed.). Harcourt Brace College Publishers.

Belle, I. (2015). *From Economic Zone to Eco-City – Urban Governance and Urban Development Trends in Tianjin's Coastal Area* (1st ed.). Borntraeger Science Publishers.

Bettencourt, L. M. (2013). The Origins of Scaling in Cities. *Science, 340*(6139), 1438–1441.

Bhattacherjee, A. (2012). *Social Science Research: Principles, Methods, and Practices.* (2nd ed.). Global Text Project. Retrieved September 22, 2018, from http://scholarcommons.usf.edu/oa_textbooks/3

Bibri, S. E. (2020a). The Eco-City and Its Core Environmental Dimension of Sustainability: Green Energy Technologies and Their Integration with Data-Drive Smart Solutions. *Energy Informatics, 2020*(3), 1–26.

Bibri, S. E. (2020b). Smart Eco-City Strategies and Solutions for Sustainability: The Cases of Royal Seaport, Stockholm and Western Harbour, Malmö, Sweden. *Urban Science, 11*(6), 1–42.

Black, A. (2006). *UK Design Council on User-Centred Design and Experience Design*. Retrieved April 19, 2018, from https://blog.experientia.com/uk-design council on user centred design and experience design/

Bluepath City Consulting. (2013). *Beautiful China: Eco-City Indicators Guidebook*. Tongxin Press.

Bonnes, M., & Secchiaroli, G. (1995). *Environmental Psychology: A Psycho-Social Introduction* (1st ed.). Sage Publications.

Bordon Area Action Group. (2009). *Government-Council Stitch-Up for Bordon*. Retrieved April 19, 2023, from http://www.baaga.co.uk

Bordon Area Action Group. (Undated). *Welcome to BAAG*. Retrieved April 19, 2023, from http://www.baaga.co.uk

Bossel, H. (1999). *Indicators for Sustainable Development: Theory, Method, Applications. A Report to the Balaton Group* (1st ed.). The International Institute for Sustainable Development (IISD).

Boyatzis, R. E. (1998). *Transforming Qualitative Information: Thematic Analysis and Code Development* (1st ed.). SAGE Publications Inc.

Brannen, J. (2008). *The Practice of a Mixed Methods Research Strategy: Personal, Professional and Project Considerations*. **Cited in:** Bergman, M. M. (2008). *Advances in Mixed Methods Research* (pp. 53–65). SAGE Publications Ltd.

Bronfenbrenner, U. (1979). *The Ecology of Human Development: Experiments by Nature and Design* (1st ed.). Harvard University Press.

Brewer, J., & Hunter, A. (2006). *The Multimethod Approach and Its Promise*. **Cited in:** Brewer, J., & Hunter, A. (2006). *Foundations of Multimethod Research* (pp. 1–14). SAGE Publications, Inc.

Braun, V., Clarke, V., Hayfield, N., & Terry, G. (2019). Thematic Analysis. **Cited in:** Braun, V *et al.* (2019). *Handbook of Research Methods in Health Social Science* (pp. 843–860).

Bryman, A. (2016). *Social Research Methods* (5th ed.). Oxford University Press.

Buckle, C. (2017). Residential Mobility and Moving Home. *Geography Compass, 11*(5), Article e12314.

Bulmer, M., et al. (1982). *Social Research Ethics: An Examination of the Merits of Covert Participant Observations* (1st ed.). Holmes and Meier.

Caprotti, F. (2013). Critical Research on Eco-Cities? A Walk Through the Sino-Singapore Tianjin Eco-City, China. *Cities, 36*(2014), 10–17.

Caprotti, F. (2014a). Critical Research on Eco-Cities? A Walk through the Sino-Singapore Tianjin Eco-City, China. *Cities, 36*, 10–17.

Caprotti, F. (2014b). Eco-Urbanism and the Eco-City or Denying the Right to the City? *Antipode, 46*(5), 1285–1303.

Caprotti, F. (2015). *Eco-Cities and the Transition to Low Carbon Economies* (1st ed.). Palgrave Macmillan.

Caprotti, F., Springer, C., & Harmer, N. (2015). 'Eco' for Whom? Envisioning Eco-Urbanism in the Sino-Singapore Tianjin Eco-city, China. *International Journal of Urban and Regional Research, 39*(3), 495–517.

Caprotti, F., & Gong, Z. Y. (2017). Social Sustainability and Residents' Experiences in a New Chinese Eco-City. *Habitat International, 61*(2017), 45–54.

Caprotti, F. (2018). Future Cities: Moving from Technical to Human Needs. *Palgrave Communications, 4*(35), 1–4.

Carr, F. S., Rivline, M., & Stone, L. (1992). *Public Space* (1st ed.). University Press.

Carvalho, L. (2015). Smart Cities from Scratch? A Socio-Technical Perspective. *Cambridge Journal of Regions, Economy and Society, 8*(1), 43–60.

Catharine, W. T. (2011). Linking Landscape and Health: The Recurring Theme. *Landscape and Urban Planning, 99*(3/4), 187–195.

Chai, Y. W., Liu, Z. L., & Zhang, Y. (2014) *Space-Time Behaviour Research and Application in China.* **Cited in:** Kwan, P., Richardson, D., Wang, D. G., & Zhou C. H. (2014). *Space-Time Integration in Geography and GIScience: Research Frontiers in the US and China* (1st ed., pp. 21–38). Springer.

Chang, I.-C.C., & Sheppard, E. (2013). China's Eco-Cities as Variegated Urban Sustainability: Dongtan Eco-City and Chongming Eco-Island. *Journal of Urban Technology, 20*(1), 57–75.

Chang, I.-C.C. (2017). Failure Matters: Reassembling Eco-Urbanism in a Globalizing China. *Environment and Planning A, 49*(8), 1719–1742.

Chang, I.-C. C. (2018). *Let's Focus on Connections: Reading Chinese Eco-Cities from a Relational, Multi-Scalar Perspective.* **Cited in:** Yu, L., & Caprotti, F. (2018). *Sustainable Cities in Asia* (1st ed., pp. 60–74). Routledge.

Chen, G. (2016). *Tianjin Eco-City: A Low-Carbon Model in Sino-Singapore Cooperation.* **Cited in:** *Singapore-China Relations: 50 Years* (pp. 131–152). World Scientific.

Chen, F. (2016b). The Deign Dimensions of China's Planning System: Urban Design for development Control. *International Planning Studies, 21*(1), 81–100.

Cheng, I. (2022, February 22). *New Urbanism Perspective on Urban Planning of Singapore.* Retrieved May 14, 2023, from https://urbandesignlab.in/new-urbanism-perspective-on-urban-planning-of-singapore/

China Academy of Science. (2020). *Big Earth Data in Support of the Sustainable Development Goals.* Science Press and EDP Sciences.

Chokor, B. A. (1987). Research Policy and Review 13 Environmental-Behaviour-Design Research Techniques: An Appraisal and Review of the Literature with Special Reference to Environmental and Planning-Related Information Needs in the Third World. *Environment and Planning A, 19*(1), 7–32.

Churchman, A. (1993). *A Differentiated Perspective on Urban Quality of Life: Women, Children and Elderly.* **Cited in:** Bonnes, M. (1993). *Perception and Evaluation of Environmental Quality* (1st ed., pp. 165 178). UNESCO Programme on Man and Biosphere.

Churchman, A. (2002). *Environmental Psychology and Urban Planning: Where Can the Twain Meet?* Retrieved December 10, 2022, from https://www.researchgate.net/publication/284027102

Clark, V., & Braun, V. (2017). Thematic Analysis. *The Journal of Positive Psychology, 12*(3), 297–298.

Corner, J., Mertins, D., & Najle, C. (2003). *Landscape Urbanism: A Manual for the Machinic Landscape* (1st ed.). Architectural Association Press.

Cowley, R. (2016). *Science Fiction and the Smart Eco-City.* The Society for the History of Technology Annual Meeting 2016. Singapore.

Cresswell, J. W. (1994). *Research Design: Qualitative and Quantitative Approaches* (1st ed.). SAGE Publications.

Cresswell, T. (2006). *On the Move: Mobilities in the Modern Western World* (1st ed.). Routledge.

Cresswell, T. (2015). *Place: An Introduction* (2nd ed.). Wiley Blackwell.

Cullen, G. (1961). *The Concise Townscape* (1st ed.). Routledge.

Cugurullo, F. (2018). Exposing Smart Cities and Eco-Cities: Frankenstein Urbanism and the Sustainability Challenges of the Experimental City. *Environment and Planning a: Economic Space, 50*(1), 73–92.

Darley, G. (1975). *Villages of Vision: A Study of Strange Utopias* (1st ed.). Architectural Press.

David, M., & Sutton, C. D. (2004). *Social Research: The Basics* (1st ed.). Sage.

Deng, W., & Cheshmehzangi, A. (2018). *Eco-Development in China: Cities, Communities and Buildings* (1st ed.). Palgrave Macmillan.

Dewar, D., & Watson, V. (1990). *Urban Markets: Developing Informal Retailing* (1st ed.). Routledge.

DeWalt, K. M., & DeWalt, B. R. (2002). *Participant Observation: A Guide for Fieldworkers* (2nd ed.). AltaMira Press.

Dobbins, M. (2009). *Urban Design and People* (1st ed.). John Wiley & Sons Inc.

Douglas, I., et al. (2012). *The Routledge Handbook of Urban Ecology* (1st ed.). New York.

Easthope, H. (2009). Fixed Identities in a Mobile World? The Relationship between Mobility, Place and Identity. *Identities: Global Studies in Power and Culture, 16*, 61–82.

Elster, J. (2015). *Explaining Social Behaviour: More Nuts and Bolts for the Social Sciences* (2nd ed.). Cambridge University Press.

Fairclough, N. (1989a). *Language and Power*. Addison-Wesley Longman.

Fainstein, S. S. (2010). *Social Justice and Ecological Urbanism.* **Cited in:** Mostafavi, M., & Doherty, G. (2010.) *Ecological Urbanism* (1st ed., pp. 300–301). E&B Engelhardt und Bauer.

Fathi, S. (2012). The Analysis of Social Relations in Sustainable Urban Space. *Journal of Iranian Social Development Studies, 4*, 47–64.

Federico, C. (2014). Critical Research on Eco-City? A Walk through the Sino-Singapore Tianjin Eco-City, China. *Cities, 36*(2), 10–17.

Federico, C., et al. (2015). Eco for Whom? Envisioning Eco-Urbanism in the Sino-Singapore Tianjin Eco-City, China. *International Journal of Urban and Regional Research, 39*(3), 495–517.

Flyvbjerg, B. (2006). Five Misunderstandings about Case-Study Research. *Qualitative Inquiry, 12*(2), 219–245.

Fossey, E., Harvey, C., McDermott, F., & Davidson, L. (2002). Understanding and Evaluating Qualitative Research. *Australian and New Zealand Journal of Psychiatry, 36*(6), 717–732.

Francis, M. *Urban Open Space: Designing for User Needs* (1st ed.). Island Press.

Friedmann, J., & Hudson, B. (1974). Knowledge and Action: A Guide to Planning Theory. *Journal of the American Institute of Planners, 40*(1), 2–16.

Fujita, M., & Krugman, P. (1995). When is the Economy Monocentric? *Regional Science and Urban Economics, 25*(4), 505–528.

Fuhrer, U., Kaiser, F. G., & Hartig, T. (1993). Place Attachment and Mobility during Leisure Time. *Journal of Environment Psychology, 13*, 309–321.

Fuller, B., & Romer, P. (2014). Urbanization as Opportunity. *World Bank Policy Research Working Paper No.6874.* Retrieved May 20, 2019, from https://papers.ssrn.com/sol3/papers.cfm?abstract_id=2439696

Glaeser, E. L. (2012). *Triumph of the City: How Our Greatest Invention Makes Us Richer, Smarter, Greener, Healthier and Happier* (1st ed.). Penguin Press.

Gibbs, G. R. (2007). Thematic Coding and Categorizing. *Analysing Qualitative Data, 703*, 38–56.

Gibson, E. J., & Pick, A. D. (2000). *An Ecological Approach to Perceptual Learning and Development* (1st ed.). Oxford University.

Gibson, J. J. (1982). *Notes on Affordances.* **Cited in:** Reed, E., & Jones, R. (1982). *Reasons for Realism: The Selected Essays of James J. Gibson* (1st ed., pp. 401–418). Lawrence Erlbaum Associates Inc.

Gibson, J. J. (1986). *The Ecological Approach to Visual Perception* (1st ed.). Lawrence Erlbaum Associates Inc.

Griffith, M. (2013). Deep Ecological Urbanism: Ecology, Efficiency and Ethics in Urban Design. Retrieved January 11, 2018, from https://treehuggingurbanism.files.wordpress.com/2013/05/griffith_deepecologicalurbanism_spreads1.pdf

Gaffron, P., Huismans, G., & Skala, F. (2005). Eco-City: A Better Place to Live. Deliverable of the Project Eco-City 'Urban Development towards Appropriate

Structures for Sustainable Transport (2002–2005). Retrieved March 16, 2019, from https://www.oekostadt.at/root/img/pool/files/book_1.pdf

Garcia Ramon, M. D., Ortiz, A., & Prats, M. (2004). Urban Planning, Gender and the Use of Public Space in a Peripheral Neighbourhood of Barcelona. *Cities, 21*, 215–223.

Gattegno, N. (2011). Ecological Urbanism. *Architectural Theory Review, 16*(1), 73–75.

Gažová, D. (2011). Urban-Landscape Contexts of Creation and Planning of Urbanized Landscape (in Slovak). *Urbanita, 23*(4), 8–13.

Gehl, J. (1987). *Life between Buildings: Using Public Space* (2nd ed.). Van Nostrand Reinhold.

Gehl, J. (2010). *Cities for People* (1st ed.). Island Press.

Gehl Architects Company. (2015). *Public Life Diversity Toolkit: A Prototype for Measuring Social Mixing and Economic Integration in Public Space.* Retrieved August 11, 2020, from https://gehlinstitute.org/wp-content/uploads/2017/02/Gehl_PublicLifeDiversityToolkit_Pages-1.pdf

Geroe, S. J. W. (2017). The Sino-Singapore Tianjin Eco-City: A Case Study of Chinese Experimental Regulatory and Institutional Development. *International Journal of Sustainable Development and Planning, 12*(6), 987–994.

Gifford, R. (2002). *Environmental Psychology: Principles and Practice* (1st ed.). Optimal Books Publishers.

Gifford, R. (2007). *Environmental Psychology: Principles and Practices* (4th ed.). Optimal Book.

Gill, J., et al. (2008). Methods of Data Collection in Qualitative Research: Interviews and Focus Groups. *British Dental Journal, 204*, 291–295.

Good, R. M., Rivero, J. J., Zitcer, A., Umemoto, K., Lake, R. W., Baum, H., Forester, J., & Harrison, P. (2017). Confronting the Challenge of Humanist Planning/Towards a Humanist Planning/A Humanist Perspective on Knowledge for Planning: Implications for Theory, Research and Practice./ To Learn to Plan, Write Stories/Three Practices of Humanism and Critical Pragmatism./Humanism or Beyond *Planning Theory & Practice*, *18*(2), 291–319.

Google Map. (27th March 2022). *Map of the SSTEC.* Retrieved March 27, 2022, from https://www.google.co.uk/maps/place/Sinosingapore+Tianjin+ECO+City+Service+Centre/@39.1154217,117.7499684,3173m/data=!3m1!1e3!4m5!3m4!1s0x35f2782eb75eb85f:0xd57b86dcf3e668ae!8m2!3d39.1242241!4d117.743853

Griffith, M. (2013). *Ecological Urbanism.* Retrieved May 11, 2021, from https://treehuggingurbanism.wordpress.com/wpcontent/uploads/2013/05/griffith_deepecologicalurbanism_spreads1.pdf

Gudmundsson, H., Hall, R. P., & Marsden, G. (2016). *Japan's Eco-Model City Program*. **Cited in**: Gudmundsson, H. *et al.* (2016). *Sustainable Transportation: Indicators, Frameworks and Performance Management* (1st ed., pp. 277–295). Springer Berlin.

Guion, L. A. (2002). *Triangulation: Establishing the Validity of Qualitative Studies*. Publication FCS6014. Gainsville, FL: Institute of Food and Agricultural Sciences, University of Florida.

Habibi, S. M. (2010). *From Town to City: A Historical Analysis of the Meaning of the City and Its Form*. Tehran University Press.

Hagan, S. (2015). *Ecological Urbanism: The Nature of the City* (1st ed.). Routledge.

Hall, E. T. (1976). *Beyond Culture* (1st ed.). Anchor Books.

Hamel, J., Dufour, S., & Fortin, D. (1993). *Case Study Methods* (1st ed.). Newbury Park.

Harding, S. (2015). *Objectivity and Diversity: Another Logic of Scientific Research* (1st ed.). University of Chicago Press.

Hardy, D. (1989). *The Garden City Campaign: An Overview*. **Cited in:** Ward, S. V. (1992). *The Garden City: Past, Present and Future* (pp. 187–209). Spon, E & FN.

Hartig, T. (2007). *Three Steps to Understanding Restorative Environments as Health Resource*. **Cited in:** Hartig, T. (2007). *Open Space: People Space* (1st ed.). Taylor and Francis.

Haynes, R. (2008). Impact of Workplace Connectivity on Office Productivity. *Journal of Corporate Real Estate, 10*(4), 286–302.

He, S. J., & Qian, J. X. (2017a). From an Emerging Market to a Multifaceted Urban Society: Urban China Studies. *Urban Studies, 54*(4), 827–846.

Heimsath, C. (1978). Behavioural Architecture: Design as if People Mattered. *Cornell Hospitality Quarterly, 19*(1), 40–48.

Hodson, M., & Marvin, S. (2010). Urbanism in the Anthropocene: Ecological Urbanism or Premium Ecological Enclaves? *Cities, 14*(3), 298–313.

Holmstedt, L., Brandt, N., & Robert, K. H. (2017). Can Stockholm Royal Seaport be Part of the Puzzle towards Global Sustainability? From Local to Global Sustainability Using the Same Set of Criteria. *Journal of Cleaner Production, 140*(1), 72–80.

Howard, E. (1902). *Garden Cities of To-morrow* (2nd ed.). S. Sonnenschein Co., Ltd.

Hu, M. C., Wu, C. Y., & Shih, T. (2016a). Creating A New Socio-Technical Regime in China: 10 Evidence from the Sino-Singapore Tianjin Eco-City. *Futures, 70*(June), 1–12.

Hu, M. C., Wadin, J. L., Lo, H. C., & Huang, J. Y. (2016b). Transformation towards an Eco-City: Lessons from Three Asian Cities. *Journal of Cleaner Production, 12*(June), 77–87.

Huang, M., Wang, Y., Wang, C., & Zhang, J. (2020). Human-Oriented Life Circle Facilities Configuration ("生活圈"之辨—基于"以人为本"理念的生活圈设置配置探讨). *The Planner, 2020*(36), 79–85.

Hugel, S. (2017). From Garden City to Smart City. *Urban Planning, 2*(3), 1–4.

Hult, A. (2013). Swedish Production of Sustainable Urban Imaginaries in China. *Journal of Urban Technology, 20*(1), 77–94.

International New Town Institution. (2017). *Tianjin Eco-City, China, Asia.* Retrieved March 27, 2022, from http://www.newtowninstitute.org/newtowndata/newtown.php?newtownId=1664

Jabareen, Y. R. (2006). Sustainable Urban Forms: Their Typologies, Models and Concepts. *Journal of Planning Education and Research, 26*(1), 38–52.

Jacobs, A., & Appleyard, D. (2007). Toward an Urban Design Manifesto. *Journal of the American Planning Association, 53*(1), 112–120.

Jacobs, J. (1961). *The Death and Life of Great American Cities* (1st ed.). Random House.

Jacobs, J. (1985). *Downtown is for People.* **Cited in:** Willian, H., & Whyte, J. R. (1985). *The Exploding Metropolis* (1st ed.). University of California Press.

JETRO. (Undated). *13 Eco-Model Cities of Japan.* Retrieved February 15, 2019, from http://www.jetro.org/content/807/#cities

Jin, B. J. (2017). *Sino-Singapore Eco-City: Challenges between A Ghost Town and A Sleeping Town (Chinese Version).* 21st Century Business Herald. Retrieved April 15, 2023, from http://business.sohu.com/20130711/n381251056.shtml

Johnson, R., et al. (2007). Toward a Definition of Mixed Methods Research. *Journal of Mixed Methods Research, 1*(2), 112–133.

Jong, M. D., et al. (2016). Eco-City Development in China: Addressing the Policy Implementation Challenge. *Journal of Cleaner Production, 134*(2016), 31–41.

Jordan, P. (2000). *Designing Pleasurable Products: An Introduction to the New Human Factors* (1st ed.). Taylor and Francis.

Jones, A. (1996). The Psychology of Sustainability: What Planners can Learn from Attitude Research. *Journal of Planning Education and Research.*

Joss, S. (2009). Eco-cities: A Global Survey 2009. *WIT Transactions on Ecology and the Environment, 129*(May), 239–250.

Joss, S. (2010). Eco-Cities – A Global Survey 2009. *WIT Transactions on Ecology and The Environment.*

Joss, S. (2011). Eco-Cities: The Mainstreaming of Urban Sustainability: Key Characteristics and Driving Factors. *International Journal of Sustainable Development Planning, 6*(2), 1–18.

Joss, S., Tomozeiu, D., & Cowley, R. (2012). Eco-City Indictors: Governance Challenges. *WIT Transactions on Ecology and the Environment, 155*(12), 109–120.

Joss, S., Cowley, R., & Tomozeiu, D. (2013). Towards the 'Ubiquitous Eco-City': An Analysis of the Internationalisation of Eco-City Policy and Practice. *Urban Research and Practice, 6*(1), 54–74.

Joss, S., & Molella, A. P. (2013). The Eco-City as Urban Technology: Perspective on Caofeidian International Eco-City, China. *Journal of Urban Technology, 20*(1), 115–137.

Joss, S., Cowley, R., de Jong, M., Müller, B., Park, B.-S., Rees, W., Roseland, M., & Rydin, Y. (2015). *Tomorrow's City Today: Prospects for Standardising Sustainable Urban Development* (1st ed.). University of Westminster.

Joss, S., David, H., & Serra, L. (2022). Eco-Neighbourhood and the Question of Locational Advantage: A Socio-Spatial Analysis of French 'ÉcoQuartiers.' *Cities, 126*, Article 103643.

Kates, R., Parris, T., & Leiserowitz, A. (2005). What is Sustainable Development? Goals, Indicators, Values, and Practice? *Environment, 47*(3), 8–21.

Keeton, R. (2011). *Rising in the East: Contemporary New Towns in Asia* (1st ed.). SUN Architecture Publishers.

Kidder, L., & Fine, M. (1987). Qualitative and Quantitative Methods: When Stories Converge. *New Directions for Program Evaluation, 35*, 57–76.

Klinenberg, E. (2018). *Palaces for the People: How Social Infrastructure Can Help Fight Inequality, Polarization and the Decline of Civic Life* (1st ed.). Penguin.

Kurniaty, R. (2014). Local Elites and Public Space Sustainability: The Local Elite Roles in the Presence and Usage of Public Space in Malang Raya, Indonesia. *Environmental Science, 20*, 506–515.

Kvale, S., & Brinkmann, S. (2009). *Interview: Learning the Craft of Qualitative Research Interviewing* (1st ed.). SAGE Publications.

Lake, R. W. (2017). A Humanist Perspective on Knowledge for Planning: Implications for Theory Research and Practice. *Planning Theory & Practice, 18*(2), 302–305.

Lang, J. (2005). *Urban Design: A Typology of Procedures and Products* (1st ed.). Architectural Press.

Lang, J., Burnette, C., Moleski, W., & Vachon, D. (1974). *Designing for Human Behaviour: Architecture and the Behavioural Sciences* (1st ed.). John Wiley & Sons Inc.

Latham, A., & Layton, J. (2019). Social Infrastructure and the Pubic Life of Cities: Studying Urban Sociality and Public Spaces. *Geography Compass, 13*(7), 1–15.

Lawlor, E., Tasker, M., Lee, R., & Hay, S. (2018). *The Pedestrian Pound: The Business Case for Better Streets and Places.*

Lawson, B. (2001). *The Language of Space* (1st ed.). Reed Educational and Professional Publishing Ltd.

Lehmann, S. (2011). *The Principle of Green Urbanism: Transforming the City for Sustainability* (1st ed.). Earthscan.

Li, C. H., & Feng, C. Y. (2012). The Application of Community Model in the Residential Planning of the SSTEC. (邻里模式在中新天津生态城住区规划中的应用). *Cities*, (3), 51–54.

Lin, Z. J. (2018). Ecological Urbanism in East Asia: A Comparative Assessment of Two Eco-Cities in Japan and China. *Landscape and Urban Planning, 179*, 90–102.

Liu, L. (2009). A Regional Divide in Sustainability Performance in China: Geography, Governance and Income. *Sustainable Development, 19*(4), 250–262.

Liu, S. J., & Zhang, Y. S. (2017). *A New Era of Green Development: China's Green Transformation 2050.* Report from China Council for International Cooperation on Environment and Development.

Liu, X., et al. (2020). Urban Form, Shrinking Cities and Residential Carbon Emissions: Evidence from Chinese City-Regions. *Applied Energy, 261*, 1–12.

Lobo, J. M. *et al.* (2020). Urban Science: Integrated Theory from the First Cities to Sustainable Metropolises. Retrieved January 11, 2023, from https://papers.ssrn.com/sol3/papers.cfm?abstract_id=3526940

Lye, L. F., & Chen, G. (2010). *Towards a Liveable and Sustainable Urban Environment* (1st ed.). World Scientific Publishing Co., Ltd.

Lyer, N. K. (1995). *Towards Responsive Environments: A Case for Urban Design and Participation.*

Lynch, K. (1960). *The Image of the City* (1st ed.). MIT Press.

Lynch, K. (1972). *The Openness of Open Space* (1st ed.). Arts of Environment.

Lynch, K. (1984). *Good City Form* (1st ed.). The MIT Press.

Lu, D. D., & Chen, M. X. (2015). Several Viewpoints on the Background of Compiling the National New-Type Urbanization XE "New-Type Urbanization" Planning (2014–2020). *Acta Geographica Sinica, 2015*(2), 179–185.

Lu, M. (2017). *The Power of Space: Geography, Politics and Urban Development (空间的力量:地理、政治与城市发展)* (2nd ed.). Shanghai People's Publishing House.

Manzo, L. C. (2005). For Better or Worse. Exploring Multiple Dimensions of Place Meaning. *Journal of Environmental Psychology, 25*(1), 67–86.

Marshall, C., & Rossman, G. B. (2015). *Designing Qualitative Research* (6th ed.). Sage Publications Inc.

Marshall, P. L. (1992). Research Ethic in Applied Anthropology. *Ethic and Human Research, 14*(6), 1–5.

Maier, J. R. A., & Fadel, G. M. (2009). Affordance Based Design: A Relational Theory for Design. *Research in Engineering Design, 20*(1), 13–27.

Marinic, G. (2012). Book Review. *Design Issues, 28*(1), 105–106.

Maslow, A. H. (1943). A Theory of Human Motivation. *Psychological Review, 50*(4), 370–396.

Massey, D. (1994). *A Global Sense of Place. Cited. In: Massey, D. (1994). Space, Place and Gender* (1st ed.). Blackwell.

Masso, A. D. (2019). Between Fixities and Flows: Navigating Place Attachments in an Increasingly Mobile World. *Journal of Environmental Psychology, 61*(2019), 125–133.

Matan, A., & Newman, P. (2016). *People City: The Life and Legacy of Jan Gehl* (1st ed.). Island Press.

May, K. A. (1991). *Interview Techniques in Qualitative Research: Concerns and Challenges.* **Cited in**: Morse, J. M. *Qualitative Nursing Research: A Contemporary Dialogue* (2st ed., pp. 188–201). Sage Publications, Inc.

McHarg, I. L. (1967). An Ecological Method for Landscape Architecture. *Landscape Architecture, 57*(1), 105–107.

MEEDDAT. (2008). *ÉcoCités: Cities of Sustainable Development: 15 Major Architectural, Social and Energy Innovation Projects at the City Level.* Ministry of Ecology, Energy, Sustainable Development and Regional Planning. Retrieved March 12, 2019, from https://www.ecologie.gouv.fr/

Ministry of National Development, Singapore Government Agency Website. (2021, August 19). *Sino-Singapore Tianjin Eco-City: A Model for Sustainable Development.* Retrieved April 10, 2023, from https://www.mnd.gov.sg/tianjinecocity/collaboration

Mostafavi, M., & Doherty, G. (2010). *Ecological Urbanism* (1st ed.). E&B engelhardt und bauer.

Mumford, L. (1968). *The City in History: Its Origins, Its Transformations, and Its Prospects* (1st ed.). Mariner Books.

Murrain, P. (1993). *Urban Expansion: Look Back and Learn.* **Cited in**: Hayward, R., & McGlynn. (1993). *Making Better Places: Urban Design Now* (pp. 83–94). Oxford.

Najafi, T. F., et al. (2016). Observation in Grounded Theory and Ethnography: What are the Differences? *Iranian Red Crescent Medical Journal, 18*(11), 40786.

Nasar, J. L. (1998). *The Evaluative Image of the City* (1st ed.). Sage Publications.

National Bureau of Statistics of China (NBS). (2022). *China Statistical Yearbook 2022.* China Statistical Publishing House.

Neisser, U. (1967). *Cognitive Psychology* (1st ed.). Appleton, Century & Crofts.

Newman, O. (1973). *Defensible Space* (1st ed.). Architectural Press.

ODPM. (2005). *Planning Policy Statement 1: Delivering Sustainable Development*. HMSO.

Ong, C. X. Y. (2019). *The Making of an Eco-City: An Examination of the Sino-Singapore Tianjin Eco-City as a New Model of Transnational New Town Development.* Retrieved March 27, 2022, from https://dspace.mit.edu/handle/1721.1/123937

Parham, S. (2013). *Shaping Sustainable Urbanism: Are Garden Cities Answer?* Retrieved January 20, 2018, from http://uh-sustainable.co.uk/docs/Shaping_sustainable_urbanism.pdf

Parsons, K. C., & Schuyler, D. (2002). *From the Garden City to Green Cities: The Legacy of Ebenezer Howard* (1st ed.). Johns Hopkins University Press.

Patton, M. Q. (2015). *Qualitative Evaluation and Research Methods: Integrating Theory and Practice* (4th ed.). Sage Publications.

Pakzad, J. (2007). *Development of Thoughts in Urbanization: From Quantity to Quality* (1st ed.). New Towns Civil Company.

Pepper, D. (1996). *Modern Environmentalism: An Introduction* (1st ed.). Routledge.

Perin, C. (1970). *With Man in Mind: An Interdisciplinary Prospectus for Environmental Design* (1st ed.). MIT Press.

Perovic, S., & Kurtovic, F. N. (2012). Visual Perception of Public Open Spaces in Niksic. *Social and Behavioural Sciences, 68*, 921–933.

Pickerill, J. (2013). *Building Liveable Cities: Urban Low Impact Development as Low Carbon Solution?* **Cited in**: Bulkeley, H., Broto, V. C., Hodson, M., & Marvin, S. (2013). *Cities and Low Carbon Transitions* (1st ed., pp. 178–197). Routledge.

Pow, C. P., & Neo, H. (2010.) *Building Ecotopia: Critical Reflections on Eco-City Development in China* (pp. 91–105). **Cited in**: Lye, L. F., & Chen, G. *Towards a Liveable and Sustainable Environment: Eco-Cities in East Asia*. World Scientific Publishing.

Prior, L. (2008). Repositioning Documents in Social Research. *Sociology, 42*(5), 821–836.

Pugh, C. (1996). *Sustainability and Sustainable Cities.* **Cited in**: Pugh, C. *Sustainability, Environment and Urbanization* (1st ed., pp. 135–177). Earthscan.

Punter, J. (1991). Participation in the Design of Urban Space. *Landscape Design, 200*, 24–27.

Punter, J. V., & Carmona, M. (1997). *The Design Dimension of Planning: Theory, Content and Best Practice for Design Policies* (1st ed.). Spon Press.

Raipat, V. (2021). *Human Behaviour Induced by Spatial Order.* **Cited in:** Levine, M. P. (2021). *Interpersonal Relationship* (1st ed.).

Rafieian, M., & Khodaei, Z. (2009). Evaluation Criteria and Standard Affecting the Satisfaction of Citizens in Urban Public Spaces. *Journal of Rahbord University, 53*, 227–248.

Rafieian, M., Azimi, M., & Moghaddam, H. (2021). Quality Evaluation of Urban Zones in Success of Public Spaces by the Use of Veroni Weighted Diagram (VWD): A Case Study of Lar Old City. *Amayesh Journal, 5*, 33–49.

Ranhagen, U., & Groth, K. (2012). *The SymbioCity Approach – A Conceptual Framework for Sustainable Urban Development*. SKL International.

Rapoport, E. (2010). *Defining the sustainable city: the discourses and convictions behind the eco-city.* Conference paper presented at the Summer Symposium on Sustainable Systems, Finland. Retrieved February 15, 2019, from http://engd-usar.cege.ucl.ac.uk/project/view/idprojects/9#outputs

Rapoport, E. (2015). Globalising Sustainable Urbanism: The Role of International Master Planners. *Area, 47*(2), 110–115.

Register, R. (1987). *Eco-city Berkeley: Building Cities for a Healthy Future* (1st ed.). North Atlantic Books.

Reed, E. S. (1996). *Encountering the World: Toward an Ecological Psychology* (1st ed.). Oxford University Press.

Research Team of Tianjin Eco-City KPI. (2009). *Piloting the Eco-City: Implementing the KPI System for the SSTEC (导航生态城市:天津生态城指标体系实施模式)* (1st ed.). China Architecture and Building Press.

Robert, C. (2016). *Eco-Cities: Technological Showcases or Public Spaces?* University of Westminster.

Robson, C. (2002). *Real World Research: A Resource for Social Scientists and Practitioner-Researchers* (2nd ed.). Blackwell.

Roseland, M. (1997). Dimensions of the Eco-City. *Cities, 14*(4), 197–202.

Rosenau, H. (2007). *The Ideal City: Its Architectural Evolution in Europe* (3rd ed.). Oxford.

Rogers, R. (1998). *Cities for a Small Planet* (1st ed.). Westview Press.

Rubin, I., & Elder, J. (1980). *Building for People: Behavioural Research Approaches and Directions* (1st ed.). Supt. of Docs.

Rueda, S. (2016). *Ecological Urbanism.* Urban Ecological Agency of Barcelona. Retrieved October 2, 2018 https://www.vitoria-gasteiz.org/http/wb021/contenidosEstaticos/adjuntos/es/80/86/38086.pdf

Sarangi, S., & Roberts, C. (1999). *The Dynamics of Interactional and Institutional Orders in Work-Related Setting.* **Cited in:** Sarangi, S., & Roberts, C. (1999). *Talk, Work and Institutional Order: Discourse in Medical, Mediation and Management Setting* (pp. 1–57).

Sanei, M., Khodadad, M., & Ghadim, F. P. (2017). Effective Instructions in Design Process of Urban Public Spaces to Promote Sustainable Development.

Saunders, M. N., *et al.* (2019). *Understanding Research Philosophy and Approaches to Theory Development.* **Cited in:** Saunders, M. N., *et al.* (2015). *Research Methods for Business Students* (8th ed.). Pearson Education Limited.

Schatzman, L., & Strauss, A. (1973). *Field Research: Strategies for a Natural Sociology* (1st ed.). Prentice Hall.

Science for Environment Policy. (2018). *In-Depth Report: Indicators for Sustainable Cities.* Bristol: Produced for the European Commission DG Environment by the Science Communication Unit, UWE.

Scott, J. (1990). *A Matter of Record: Documentary Sources in Social Research* (1st ed.). Polity Press.

Scopelliti, M., et al. (2015). Go Greener, Feel Better? The Positive Effects of Biodiversity on the Well-Being of Individuals Visiting Urban and Peri-Urban Green Areas. *Landscape and Urban Planning, 134*, 221–228.

Sharifi, A. (2016). From Garden City to Eco-Urbanism: The Quest for Sustainable Neighbourhood. *Sustainable Cities and Society, 20*, 1–16.

Shwayri, S. (2013). A Model Korean Ubiquitous Eco-City? The Politics of Making Songdo. *Journal of Urban Technology, 20*(1), 39–55.

Singapore Ministry of National Development Website. (2021, April 15). *What is the Concept of the SSTEC?* Retrieved June 5, 2021, from https://www.mnd.gov.sg/faqs/general/sino-singapore-tianjin-eco-city

Spaargaren, G., & Mol, A. P. J. (1992). Sociology, Environment and Modernity: Ecological Modernization as a theory of Social Change. *Society and Natural Resources, 5*(4), 323–344.

Spirn, A. W. (2000). *Ian McHarg, Landscape Architecture and Environmentalism: Ideas and Methods in Context.* **Cited in:** M. Conan. ed. (2000). *Environmentalism and Landscape Architecture* (pp. 97–114). Dumbarton Oaks.

Spirn, A. W. (2014). *Ecological Urbanism: A Framework for the Design of Resilient Cities.* **Cited in:** Foster, O. N. (2014). *The Ecological Design and Planning Reader* (1st ed., pp. 557–596). Island Press.

Stake, R. E. (2008). *Qualitative Case Studies.* **Cited in**: Denzin, N. K., & Lincoln, Y. S. (2008). *Strategies of Qualitative Inquiry* (3rd ed., pp. 119–149). Sage Publications, Inc.

SSTEC Investment and Development Co., Ltd Website. (No date shown.) *Masterplan of the SSTEC.* Retrieved March 18, 2019, from http://www.tianjineco-city.com/static/web/en/singapore3.html?lang=english

SSTEC Investment and Development Co., Ltd Website. (No date shown.) *Key Principles and KPIs.* Retrieved March 18, 2019, from http://www.tianjineco-city.com/static/web/en/singapore3.html?lang=english

Sino-Singapore Tianjin Eco-City Administrative Committee, China Eco-City Academy and Centre for Liveable Cities. (2021). *A Partnership for a Smart and Sustainable Future: The Sino-Singapore Tianjin Eco-City.* Retrieved January 11, 2022, from https://www.clc.gov.sg//docs/default-source/books/a-partnership-for-a-smart-and-sustainable-future-the-china-singapore-tianjin-eco-city.pdf

Song, F., et al. (2019). Research Progress of Urban Ecological Carrying Capacity. *Environmental Ecology, 2019*(1), 53–61.

Su, H. J., & Wei, H. K. (2018). The Course, Inspiration and Prospects of China's Urbanization over the Past 40 Years of Reform and Opening-Up. *Reform, 2018*(11), 49–59.

Sun, X. F. (2010). The Preliminary Study on Eco-City Planning: Taking the Masterplan of Sino-Singapore Tianjin Eco-City as an Example (生态城市规划

初探——以中新天津生态城总体规划为例). *New Energy and Green Building.*, *38*(8), 36–40.

Suzuku, H. *et al.* (2010). *Eco2 Cities: Ecological Cities as Economic Cities*. The World Bank.

Talen, E. (2002). The Social Goal of New Urbanism. *Housing Policy Debate, 13*(1), 165–187.

Tilt, J. H., Unfried, T. M., & Roca, B. (2007). Using Objective and Subjective Measures of Neighbourhood Greenness and Accessible Destinations for Understanding Walking Trips and BMI in Seattle, Washington. *American Journal of Health Promotion, 21*(4), 371–379.

Trancik, R. (1986a). *Finding Lost Space: Theories of Urban Design* (1st ed.). Van Nostrand Reinhold Company.

Tomozeiu, D., & Joss, S. (2014). Adapting Adaptation: The English Eco-Town Initiative as Governance Process. *Ecology and Society, 19*(2), 20.

Toth, G., & Espiau, R. (2008). *Community Based Street Design, Project for Public Space*. Retrieved March 25, 2023, from https://www.pps.org/article/community-based-street-design

Town and Country Planning Association. (2011). *Re-Imagining Garden Cities for the Twenty-First Century: Benefits and Lessons in Bringing Forward Comprehensively Planned New Communities*. Retrieved April 25, 2018, from https://www.tcpa.org.uk/Handlers/Download.ashx?IDMF=20d46898-f701-4673-8b55-1f2f23843850

Town and Country Planning Association. (2013). *Creating Garden Cities and Suburbs Today: A Guide for Councils*. Retrieved January 20, 2018, from https://www.tcpa.org.uk/Handlers/Download.ashx?IDMF=728ee8cf-ef8f-4c51-bc5c-8c4bbd1eab9f

Town and Country Planning Association. (2015). *New Towns and Garden Cities – Lesson for Tomorrow. Stage 2: Lessons for Delivering a New Generation of Garden Cities*. Retrieved April 20, 2018, from https://www.tcpa.org.uk/lessons-for-tomorrow-stage-2-report

Town and Country Planning Association Website. (No date shown). *Garden City Principles*. Retrieved May 28, from https://www.tcpa.org.uk/garden-city-principles

Trancik, R. (1986b). Finding Lost Space: Theory of Urban Design. *Van Nostrand Reinhold, 74*, 32–38.

Turel, H. S., Yigit, E. M., & Altug, I. (2007). Evaluation of Elderly People's Requirements in Public Open Spaces: A Case Study in Bornova District (Izmir, Turkey). *Building and Environment, 42*, 2035–2045.

Turnbull, J. (2010). *Ecological Urbanism Review*. Retrieved February 10, 2023, from https://www.domusweb.it/en/reviews/2010/12/02/ecologicalurbanism.html

United Nations, Department of Economic and Social Affairs, Population Division. (2022). *United Nations E-Government Survey*. Retrieved January 11, 2023, from https://publicadministration.un.org/egovkb/en-us/Reports/UN-E-Government-Survey-2022

Urry, J. (2000). Mobile Sociology. *British Journal of Sociology, 51*(1), 185–203.

Vlek, C. (2000). Essential Psychology for Environmental Policy Making. *International Journal of Psychology, 35*(2), 153–167.

Wagenaar, H. (2011). A Beckon to the Makings, Workings and Doing Human Beings: The Critical Pragmatism of John Foster. *Public Administration Review, 71*(2), 293–299.

Waldheim, C. (2010a). *The Landscape Urbanism Reader* (1st ed.). Princeton Architectural Press.

Waldheim, C. (2010b). *Landscape as Urbanism: A General Theory* (1st ed.). New Jersey.

Waldheim, C. (2010). *Weak Work: Andrea Branzi's 'Weak Metropolis' and the Projective Potential of an 'Ecological Urbanism'* **Cited in:** Mostafavi, M., & Doherty, G. (2010). *Ecological Urbanism* (1st ed., pp. 114–121). E&B Engelhardt und Bauer.

Walters, P. (2005). *Knowledge in the Making: Prototyping and Human-Centred Design Practice*. Sheffield Hallam University.

Ward, C. (1993). *New Town, Home Town: The Lessons of Experience* (1st ed.). Calouste Gulbenkian Foundation.

Ward, S. V. (2016). *The Peaceful Path: Building Garden Cities and New Towns*. Hertfordshire Press.

WCED. (1987). Report of the World Commission on Environment and Development: Our Common Future. Retrieved June 30, 2022, from https://sustainabledevelopment.un.org/documents/our-common-future.pdf

Wei, H. K. (2014). *A New Road to Urbanization with Chinese Characteristics(走中国特色的新型城镇化道路)* (1st ed.). Social Sciences Academic Press.

Weiss, R. S. (1995). *Learning from Strangers: The Art and Method of Qualitative Interview Studies* (1st ed.). Simon & Schuster Inc.

Whyte, H. W. (1980a). *The Social Life of Small Urban Space* (1st ed.). Conservation Foundation.

Whyte, W. H. (1980b). *The Social Life of Small Urban Spaces* (1st ed.). The Conservation Foundation.

William, D. R., & McIntyre, N. (2012). *Place Affinities, Lifestyle Mobilities and Quality of Life*. **Cited in:** William, D. R., & McIntyre, N. (2012). *Handbook of Tourism and Quality of Life Research* (1st ed., pp. 209–231). Springer.

William, D. R. (2014). Making Sense of Place: Reflection on Pluralism and Positionality in Place Research. *Landscape and Urban Planning, 131*, 74–82.

William, D. R. (2018). *Spacing Conservation Practice: Place-Making, Social Learning and Adaptive Governance in Natural Resource Management*. **Cited**

**in:** Marsden T. (2018). *The SAGE Handbook of Nature* (Vol 3., 1st ed., pp. 285–303). Sage.

Williams, M. (2000). Interpretivism and Generalisation. *Sociology, 34*(2), 209–224.

Withagen, R., & Michaels, C. F. (2005). On Ecological Conceptualizations of Perceptual Systems and Action Systems. *Theory and Psychology, 15*, 603–620.

Withagen, R., de Poel, H. J., Araujo, D., & Pepping, G. J. (2012). Affordance can Invite Behaviour: Reconsidering the relationship between Affordances and Agency. *New Ideas in Psychology, 30*(2), 250–258.

World Bank. (2009). *Sino-Singapore Tianjin Eco-City: A Case Study of an Emerging Eco-City in China (Technical Assistance Report)*. Retrieved February 17, 2019, from https://openknowledge.worldbank.org/handle/10986/28143

World Bank. (2010). *Eco2 Cities Brochure*. Retrieved February 15, 2019, from http://siteresources.worldbank.org/INTEASTASIAPACIFIC/Resources/Eco2_Cities_Brochure

World Commission on Environment and Development. (1987). *Brundtland Report: Our Common Future*. Retrieved February 15, 2023, from https://sustainabledevelopment.un.org/content/documents/5987our-common-future.pdf

Wu, F. L. (2000). The Global and Local Dimensions of Place-Making: Remaking Shanghai as A World City. *Urban Studies, 37*(8), 1359–1377.

Wu, F. L. (2007). Re-Orientation of the City Plan: Strategic Planning and Design Competition in China. *Geoforum, 38*(2), 379–392.

Wu, F. L. (2012). China's Eco-Cities. *Geoforum, 43*(2), 169–171.

Wu, F. L., Zhang, F. Z., & Wang, Z. (2015). *Planning China's Future: How Planners Contribute to Growth and Development*. RTPI Research Report. Retrieved April 24, 2022, from https://fulongwu.files.wordpress.com/2016/02/rtpi-research-report-12-planning-chinas-future-december-2015.pdf

Wu, Z. Q. (2023). *Principles of Territorial Spatial Planning (国土空间规划原理)* (1st ed.). Press of Tongji University.

Xia, Y., Zhang, C. Y., & Zhu, H. (2021). The Performance Evaluation and Difficulties of the Eco-City Construction: The Case Study of Sino-Singapore Tianjin Eco-City. (生态城建设的绩效评价及难点探源—基于中新天津生态城的案例)*Cities, 27*(14), 27–40.

Xie, L. J., et al. (2020). Urban Entrepreneurialism and Sustainable Development: A Comparative Analysis of Chinese Eco-Developments. *Journal of Urban Technology, 27*(1), 3–26.

Yigitcanlar, T., & Lee, S.-H. (2014). Korean Ubiquitous-Eco-City: A Smart-Sustainable Urban Form or A Branding Hoax? *Technological Forecasting & Social Change, 89*, 100–114.

Yilmaz, S., et al. (2017). Determining the Affordance Provided by Urban Open Spaces to Different Age Groups. *Gazi University Journal of Science Part B: Art Humanities Design and Planning, 5*(3), 1–12.
Yin, R. K. (2009). *Case Study Research: Design and Methods* (4th ed.). Safe Publications.
Yin, R. K. (2014). *Case Study Research: Design and Methods* (5th ed.). Safe Publications.
Yip, S. C. T. (2008). Eco-city: From Concept to Implementation-Case Studies in Shanghai Dongtan and Beijing Changxindian. *China City Planning Review, 2008*(4), 45–53.
Yu, K. J. (2010). *Introduction of Book Ecological Urbanism (Chinese Vision).* **Cited in:** Mostafavi, M., & Doherty, G. (2010). *Ecological Urbanism* (1st ed., pp. 2–3). E&B Engelhardt und Bauer.
Yu, L. (2014). Low Carbon Eco-City: New Approach for Chinese Urbanisation. *Habitat International, 44*(2014), 102–110.
Yurzinov, I. L. (2014). Eco-Town: Current Status and Prospects. *Economy. Taxes. Right, 6*, 71–73.
Zeisel, J. (1974). *Sociology and User Analysis Research* (1st ed.). Russell Sage Foundation.
Zhan, C. J., Jong, M. D., & Bruijn, H. D. (2018). Funding Sustainable Cities: A Comparative Study of Sino-Singapore Tianjin Eco-City and Shenzhen International Low-Carbon City. *Sustainability, 10*(4256), 1–15.
Zhang, T., & Dong, H. (2008). *Human-Centred Design: An Emergent Conceptual Model.* In Royal College of Art, April 8–10 2009. London.
Zhang, Y. S. (2021). China's Green Urbanization in the Perspective of Ecological Civilization. *Chinese Journal of Urban and Environmental Studies, 9*(1), 1–17.
Zhao C. L. *et al.* (2012). PSPL Survey: The Evaluation Method for Quality of Public Space and Public Life – The Study on Jan Gehl's Theory and Method for Public Space Design (In Chinese). *Theory of Planning and Design, TU986*(A), 34–36.
Zuniga-Teran, A. A., et al. (2016). Designing Healthy Communities: A Walkability Analysis of LEED-ND. *Frontiers of Architectural Research, 5*(4), 433–452.

# Index







Y. Wu, *Possibilities and Limitations of Chinese Eco-City Development*, New Perspectives on Chinese Politics and Society,
https://doi.org/10.1007/978-981-95-1958-3

FSC
www.fsc.org
MIX
Papier | Fördert gute Waldnutzung
FSC® C083411

Kolibri 360 GmbH
Ferdinand-Jühlke-Straße 7
99095 Erfurt, Deutschland
produktsicherheit@kolibri360.de